T-Labs Series in Telecommunication Services

Series Editors

Sebastian Möller, Quality and Usability Lab, Technische Universität Berlin, Berlin, Germany

Axel Küpper, Telekom Innovation Laboratories, Technische Universität Berlin, Berlin, Germany

Alexander Raake, Audiovisual Technology Group, Technische Universität Ilmenau, Ilmenau, Germany

It is the aim of the Springer Series in Telecommunication Services to foster an interdisciplinary exchange of knowledge addressing all topics which are essential for developing high-quality and highly usable telecommunication services. This includes basic concepts of underlying technologies, distribution networks, architectures and platforms for service design, deployment and adaptation, as well as the users' perception of telecommunication services. By taking a vertical perspective over all these steps, we aim to provide the scientific bases for the development and continuous evaluation of innovative services which provide a better value for their users. In fact, the human-centric design of high-quality telecommunication services – the so called "quality engineering" – forms an essential topic of this series, as it will ultimately lead to better user experience and acceptance. The series is directed towards both scientists and practitioners from all related disciplines and industries.

** Indexing: books in this series are indexing in Scopus **

Saman Zadtootaghaj

Quality of Experience Modeling for Cloud Gaming Services

Springer

Preface

The *gaming* industry has been one of the largest in the entertainment markets for the past several decades and is steadily growing with the introduction of emerging technologies such as hardware video encoding and the new generation of broadband cellular networks 5G. With these advancements, a new gaming paradigm called *cloud gaming* has emerged that makes gaming possible at any time, on any device, and at any place. Cloud gaming shifts the heavy computational tasks such as rendering to the cloud resources and streams a compressed video of players' gameplay back to the client in real time. Similar to other telecommunication services, cloud gaming is prone to network and compression degradations such as blockiness, blurring, and network latency. These degradations could negatively affect the quality of experience (QoE) of users. Therefore, it is of high interest for service and network providers to measure and monitor the QoE of cloud gaming services to potentially improve the satisfaction of their customers.

The present book aims at the development of a gaming quality model to predict the gaming QoE of players that could be used for planning the network service or quality monitoring of cloud gaming services. The model is developed following a modular structure approach that keeps the different types of impairment separately. Such a modular structure allows developing a sustainable model as each component can be updated by advances in that specific research area or technology. The book describes the steps toward planning and conducting subjective tests for game streaming services, analyzing the subjective data, and the process of modeling quality judgment ratings of users with different types of approaches.

This book is written for students, researchers, and engineers involved in the QoE assessment of video game streaming services. In addition, it might be useful for a person who is working on network planning and operation.

This book is the result of studies conducted during my PhD research, which was conducted in the Quality and Usability Lab at TU Berlin and Telekom Innovation Laboratories at Deutsche Telekom (known as T-Labs) between 2016 and 2021. The scope of this book was influenced by the ITU-T standard activities at Study Group 12, particularly the G.OMG work item ("Opinion model predicting gaming quality of experience for cloud gaming services"), which resulted in an ITU-T recommendation G.1072, that author contributed between 2016 to 2020.

Berlin, Germany Saman Zadtootaghaj
January 2022

Acknowledgments

Throughout the writing of this book, I have received a great deal of support and assistance. I would first like to thank Professor Dr.-Ing. Sebastian Möller, whose expertise was invaluable in formulating the research questions and methodology.

I would particularly like to thank two of my greatest and discerning colleagues Steven Schmidt and Nabajeet Barman, with whom I had numerous discussions on the various topics and projects. Thank you both for your countless support and for not only being a great colleague in every step of my PhD research but also being a caring friend. I am forever indebted to you for your help and guidance during the past 6 years. I want to give special thanks to Babak Naderi and Saeed Shafiee Sabet for walking me through the crowdsourcing and online gaming topics.

I would like to acknowledge my colleagues from the Quality and Usability Lab for their wonderful collaboration. I always cherish the memories shared with my colleagues Jan-Niklas Voigt-Antons, Stefan Uhrig, Rafael Zequeira, Gabriel Mittag, Tanja Kojic, Salar Mohtaj, Falk Schiffner, Maija Poikela, Patrick Ehrenbrink, Thilo Michael, Andres Pinilla Palacios, and Vera Schmitt. Many thanks go to Irene Hube-Achter and Yasmin Hillebrenner for their administrative and technical support during my years at QU Lab.

My PhD journey started with the European project QoE-Net, where I collaborated with a great team of PhD students and advisers from different institutes. Among them, I would like to give a special thank you to my manager at Deutsche Telekom, Bernhard Feiten, for his kindness and endless support during all these years. I also would like to thank all mentors of the QoE-Net project, including Prof. Maria Martini, Prof. Touradj Ebrahimi, Prof. Andrew Perkis, Prof. Alexander Raake, and Prof. Lingfen Sun, as well as all researchers that I collaborated with, Werner Robitza, Alcardo Alex Barakabitze, Subbareddy Darukumalli, Arslan Ahmad, Avsar Asan, and Elisavet Grigoriou. Special thanks to Anne-Flore Perrin and Prof. Ebrahimi for their great advice and collaboration during my visit to EPFL.

My collaboration with the Audiovisual Technology Group at TU Ilmenau was always the best of its kind. I would like to thank Prof. Alexander Raake and his team, particularly my great friends Rakesh Rao Ramachandra Rao, Steve Göring, and Stephan Fremerey, for collaborating and advising me in the video quality domain.

Many thanks to Rakesh for proofreading this book and for all advice over the past few years.

I would also like to especially thank my thesis examiner, Dr. Maria Martini, for her valuable guidance throughout my studies and for reviewing most of my publications and this book.

Ultimately, my biggest thanks are directed to my family, who supported me throughout my entire life and encouraged me in every step of my life, in particular, I must acknowledge my wife and best friend, Nasim Jamshidi. Without your endless love and encouragement, I would never have been able to complete my graduate studies. I love you, and I appreciate everything that you have done for me!

I wish to express my dearest gratitude to my parents, Soraya and Ebrahim, for their endless love and continuous support throughout my entire life. I would also like to thank my sister and brother, Diman and Siamand; you have always inspired me to do my best and have supported me in every decision I have made.

Contents

Acronyms

AC	Service Acceptance
ACM	Amount of Camera Movement
ACR	Absolute Category Rating
ACR-HR	ACR with Hidden Reference
AQ	Audio Quality
B-frames	Bidirectional-Frames
BIQI	Blind Image Quality Index
BM	Block Motion estimation
BRISQUE	Blind/Referenceless Image Spatial Quality Evaluator
CBR	Constant Bitrate
CGVDS	Cloud Gaming Video Quality DataSet
CH	Challenge
CN	Controllability
CNN	Convolutional Neural Network
CO	Competency
CQP	Constant Quantization Parameter
CRF	Constant Rate Factor
CSGO	Counter-Strike: Global Offensive
DBSQE-V	Dimension-Based Subjective Quality Evaluation Method
DMOS	Differential MOS
DSR	Dynamic Super Resolution
DoF	Degrees of Freedom
ECscale	Extended Continuous scale
EC-ACR	Extended Continuous ACR
EEG	Electroencephalography
EWMA	Exponentially Weighted Moving Average
FHD	Full High Definition
FL	Flow
FR	Full-Reference
GAN	Generative Adversarial Network
GEQ	Game Experience Questionnaire

GIPS	Gaming Input Quality Scale
GoP	Group of Picture
GVSET	Gaming Video Dataset
HEVC	H.265/High Efficiency Video Coding
HUD	Heads-Up Display
HVS	Human Visual System
IF	Impairment Factor
iGEQ	in-game Game Experience Questionnaire
IM	Immersion
IP	Internet Protocol
ISO	International Organization for Standardization
ISPs	Internet Service Providers International Telecommunication Union
ITU	Telecommunication Standardization Sector
KPIs	Key Performance Indicators
KUGVD	Kingston University Gaming Video Dataset
llhq	low latency high quality
MDS	Multidimensional Scaling
MGUE	Mobile Gaming User Experience
MLR	Multiple Linear Regression
MOS	Mean Opinion Score
NA	Negative Affect
NCA	Neighborhood Components Analysis
NFLX-PD	Netflix Public Dataset Subjective Video QoE Database
NFLX-SVQD	LIVE-NFLX-II
NIMA	Neural Image Assessment
NIQE	Natural Image Quality Evaluator
NR	No-Reference
NSS	Natural Scene Statistics
NVEnc	NVIDIA Encoder
PA	Positive Affect
PReLU	Parametric Rectified Linear Unit
PC	Personal Computer
PCA	Principal Component Analysis
PLC	Packet Loss Concealment
PR	Playing Performance
PLCC	Pearson Linear Correlation Coefficient
PSNR	Peak Signal to Noise Ratio
PX	Player Experience
QoE	Quality of Experience
QoS	Quality of Service
QP	Quantization Parameters
RBF	Radial Basis Function
RD	Rate-Distortion
RE	Responsiveness
Rec.	Recommendation

RF	Random Forest
RFE	Recursive Feature Elimination
RR	Reduced-Reference
RTP	Real-time Transport Protocol
RTSP	Real Time Streaming Protocol
SA	Static Areas
SGD	Stochastic Gradient Descent
SI	Spatial Information index
SLA	Service Level Agreement
SoA	State of the Art
SR	Super-Resolution
SRCC	Spearman's Rank Correlation Coefficient
SS	Single Stimulus
SSIM	Structural Similarity
SVR	Support Vector Regression
TC	Temporal Complexity
TE	Tension
TI	Temporal Information index
UDP	Datagram Protocol
UHD	4K/Ultra-High Definition-1
VBR	Variable Bitrate
VD	Video Discontinuity
VF	Video Fragmentation
VIFP	Visual Information Fidelity-Pixel Domain
VL	Suboptimal Video Luminosity
VMAF	Video Multimethod Assessment Fusion
VP9	Video Payload type 9
VQ	Video Quality
VQEG	Video Quality Expert Group
VR	Virtual Reality
VU	Video Unclearness
WebRTC	Web Real Time Communication

Chapter 1
Introduction

1.1 Motivation

In 1952, Sandy Douglas gave birth to the first know video game named *OXO*, also known as *tic-tac-toe*. However, only in 1967, *Brown Box* allowed customers to enjoy virtual worlds in their home environments. Over 40 years later, the US video gaming market reached $60.4 billion[1] in 2020. On the contrary to other businesses, the COVID-19 pandemic had no negative influence on the growth of the gaming industry as potential consumers were even more likely to engage in gaming.

The first form of commercial video games ran on a game console. Nowadays, video games are played on multiple types of devices such as Personal Computer (PC), and mobile devices connected to various input devices (e.g., mouse, gamepads, and keyboards), as well as output devices (e.g., 4K monitors, and Head-Mounted Displays). Over time, with the advancement of technology, video games became more and more complex both in terms of design and processing power. This growth in complexity requires players to update their end devices every few years to play the latest released version of video games. One solution for this is to move the heavy processes such as the rendering to the cloud and cut the needs for having high-end hardware devices for customers. This is the initial idea behind a Cloud Gaming service.

Cloud Gaming is characterized by game content delivered from a server to a client as a video stream with game controls sent from the client to the server. The execution of the game logic, rendering of the virtual scene, and video encoding are performed at the server, while the client is responsible for video decoding and capturing of client input [1].

[1] https://www.statista.com/topics/868/video-games/.

S. Zadtootaghaj, *Quality of Experience Modeling for Cloud Gaming Services*, T-Labs Series in Telecommunication Services,
https://doi.org/10.1007/978-3-030-98249-2_1

Cloud gaming is proposed to offer more flexibility to users by allowing them to play any game anywhere and on any type of device. Apart from processing power, cloud gaming benefits users by offering platform independence, i.e. that every game can be played on any device regardless of the operating system. For game developers, it offers security to their products against piracy and promises a new market to increase their revenue.

OnLive was one of the early cloud gaming services which received lots of attention between 2012 to 2015. While the service seemed to be very promising, it failed to attract users due to high latency and low video quality which was finally shut down in 2015 after being sold to Sony. With the advancement of network and compression technologies, new cloud gaming services have emerged recently. Among others, the following services of high-tech companies joined the market: Google Stadia, Sony's PlayStation Now, NVIDIA GeForce Now, Microsoft's Project xCloud, Amazon's Luna, and Telekom's MagentaGaming etc. Yet, the biggest challenge of all cloud gaming services is to provide a satisfying quality to their end-users.

Cloud gaming as a real-time application suffers strongly from the additional delay due to video encoding and decoding, end-to-end transmission latency, as well as other network constraints such as limited bandwidth and packet loss. The Quality of Experience (QoE) of customers, which is described by the degree of delight or annoyance of the user [2], can be negatively affected by these limitations. In order to optimize user satisfaction despite these challenges, strategies for resource and network management are highly necessary. As a ground truth for these strategies, subjective user ratings for various system configurations are required. As these subjective tests are very time-consuming and costly, there is a high interest from network and service providers to create QoE prediction models.

The *aim* of the present book is to provide quality models for cloud gaming services based on the impact of impairments introduced by typical Internet Protocol (IP) networks on the quality experienced by players. Based on the use case and available information about the network and service parameters, providers will be able to select a suitable model from a series of models that are developed to be used for network planning and quality monitoring of cloud gaming services.

The proposed models are composed of two modules, video quality, and input quality, affected by three different impairment types: transmission error, lossy compression, and latency on the control stream. Each model predicts a Mean Opinion Score (MOS) of overall gaming QoE on a five-point Absolute Category Rating (ACR) scale (cf. ITU-T Rec. P.910 [3]). In addition to the overall gaming quality score, separated predictions of the video quality and input quality are available as diagnostic information that could be used to determine the cause of impairment.

All models follow the same structure, while three different types are proposed for the prediction of video quality that is impaired by video compression. Three types of video quality models are developed with different levels of access to the video stream information:

- *Network Planning models* which can be used by various stakeholders for purposes such as resource allocation and configuration of IP-network transmission settings such as the selection of resolution, framerate, and bitrate under the assumption that the network is prone to packet loss and delay.
- *Bitstream-based models* which can be used by cloud gaming and network providers to monitor the quality based on the extraction of packet header information and network parameters, i.e., end-to-end latency and packet loss rate.
- *No-Reference Signal-based models* that can be used by cloud gaming providers to predict the gaming QoE based on the access to the compressed video signal, and network parameters.

1.2 Scope of the Book

In this section, the scope of models targeted in this book is presented, and more importantly, the aspects that are not considered in the scope are listed. The developed models target cloud gaming services that perform video streaming over Real-time Transport Protocol (RTP) (over Datagram Protocol (UDP)) and which select various video encoding parameters to adapt to the network throughput, packet loss, and end-to-end delay. The impact of network and encoding distortions on various quality features perceived by a player depends strongly on the sensitivity of a game towards these degradations. In this book, three types of game classifications are described based on the game characteristics that can be used for cloud gaming quality prediction. If no information on the game class of sensitivity is available to the user of the model, the "default" *mode of operation* should be used that assumed the highest (sensitivity) game class. The "default" *mode of operation* will result in a pessimistic quality prediction for games that are not of high complexity and sensitivity.

Considering the limitations, such as limited resources and time frame for the development of models predicting gaming QoE, the scope of the work is limited to a number of important aspects relevant to the current state of the cloud gaming services. The reasons behind the selection of parameters, range of parameters, technology, participants, input and output modalities are presented in the following chapters of the book. The developed models only work in the scope and range of the parameters that the models have been developed for (cf. Chap. 3). Therefore, the following aspects are out of the scope of the proposed models:

- Predicting the influence of the game design or the motivation of users to play them.
- The gaming QoE under the influence of social factors, especially for multiplayer games where players can communicate with each other.
- Virtual reality games requiring 3D rendering devices or displayed on Head-Mounted-Displays.

- Display sizes lower or higher than 24″. The display size of 24″ is selected as a typical PC gaming monitor at the time of the research.
- Input devices other than keyboard and mouse, e.g., gamepad and touch displays.
- Traditional online gaming services requiring the execution of a game on the client device.
- Core-gamers, as they are not assumed to be the main target of cloud gaming services.
- Online gaming that no video stream is transmitting over the broadband network.

With respect to the encoding technologies considered, the models are trained based on hardware accelerator engines for video compression and H.264/MPEG-4 AVC as the choice of video compression standard. The models are trained and evaluated at video resolution up to Full High Definition (FHD), framerates up to 60 fps, and end-to-end delay up to 400 milliseconds.

However, the No-Reference signal-based models may be considered to measure the gaming QoE for other types of video codecs and encoding settings as this mode relies on the signal information and not encoding parameters.

While the focus of the described model is on cloud gaming, the video quality module might be relevant for passive gaming video streaming applications such as Twitch.tv where only the video content is streamed to passive viewers.

1.3 Structure of the Book

Chapter 2 gives an introduction to Quality and Quality of Experience in the context of cloud gaming services. In addition, it provides a short overview of QoE influencing factors of cloud gaming service, subjective test procedures, and existing gaming quality models in the literature.

Chapter 3 presents the model framework structure and describes the necessary steps before the development of gaming QoE models, including training data, data processing, and classification of video games according to their sensitivity to network and compression impairment.

In Chap. 4, the proposed video quality models that predict the quality of compressed sequences are described. Multiple models are proposed according to the level of access to streamed video information or video encoding parameters. In total, two planning models, a bitstream model, and three signal-based models are proposed in this chapter. These models can be used as standalone models to predict the video quality of the encoded video or with the gaming QoE model to predict the impairment due to compression.

Chapter 5 provides information about the development of models to predict the impairment due to transmission errors and errors on control streamflow. In addition, the core gaming QoE model is described in this section. The section ends with a short description of Telecommunication Standardization Sector of the International Telecommunication Union (ITU-T) Recommendation (Rec.) G.1072 model that author strongly contributed to, and discusses the differences as compared to the proposed framework of the book.

The performance of the proposed impairment factor models is evaluated in Chap. 6. The performance is analyzed based on multiple gaming video quality datasets as well as an interactive quality dataset. In addition, the performance evaluation of the final gaming QoE model and ITU-T Rec. G.1072 models are presented in separate sections.

Chapter 7 provides some insight into one of the important future challenges for QoE assessment of gaming streaming services which is enhancement techniques. The enhancement models, such as super-resolution techniques, introduce new types of degradation such as adversarial degradation to the encoded video stream. The advancement of video quality enhancement models raises the concern of whether the current quality models are still usable for the enhanced video stream or not.

Finally, in Chap. 8, a summary of the key contributions of the book and limitations in the development of the models are provided. Next, the possible extensions of the model are described to give an outlook for future work.

Chapter 2
Gaming Quality of Experience

2.1 Cloud Gaming System

Cloud gaming technology is evolving rapidly with the advancement of compression and network technologies employed in the system to improve the players' QoE. Regardless of the differences in technologies employed in different cloud gaming systems, the system itself can be depicted at an abstract level, as shown in Fig. 2.1. In a cloud gaming service, the server, also called a cloud, is responsible for executing the game, including running the game logic and rendering the scenes according to the player's commands. While the end user's device, also named thin-client, is responsible for receiving the streamed audio and video signals and display them on the thin-client device. In addition, the client device captures the user commands and transmits them to the cloud gaming server. Two data flow streams are required in the system, one to transmit the control commands to the server, which is called "control flow", and another to send the audiovisual signal to the client-side. All these processes are constrained by the performance of the network between the client and the server. Several constraints, such as propagation and transmission latency, lossy channel, and limited bandwidth would influence the experience of players. Due to the limited bandwidth, the audiovisual stream must be compressed before transmission over the broadband networks. Thus, on the server side, the audiovisual content is encoded and then decoded on the client-side.

For a real-time cloud gaming service, typically, the cloud gaming service providers use hardware-accelerated engines to encode the videos under presets and settings appropriate for real-time encoding. In this book, the hardware-accelerated solution provided by NVIDIA, NVIDIA Encoder (NVEnc) for the H.264/MPEG-4 AVC codec is followed.

In addition to limited bandwidth, a cloud gaming service runs over UDP that is prone to packet loss due to network congestion and forward error correction. Due to the packet loss, re-transmission might be an option that introduces additional

S. Zadtootaghaj, *Quality of Experience Modeling for Cloud Gaming Services*,
T-Labs Series in Telecommunication Services,
https://doi.org/10.1007/978-3-030-98249-2_2

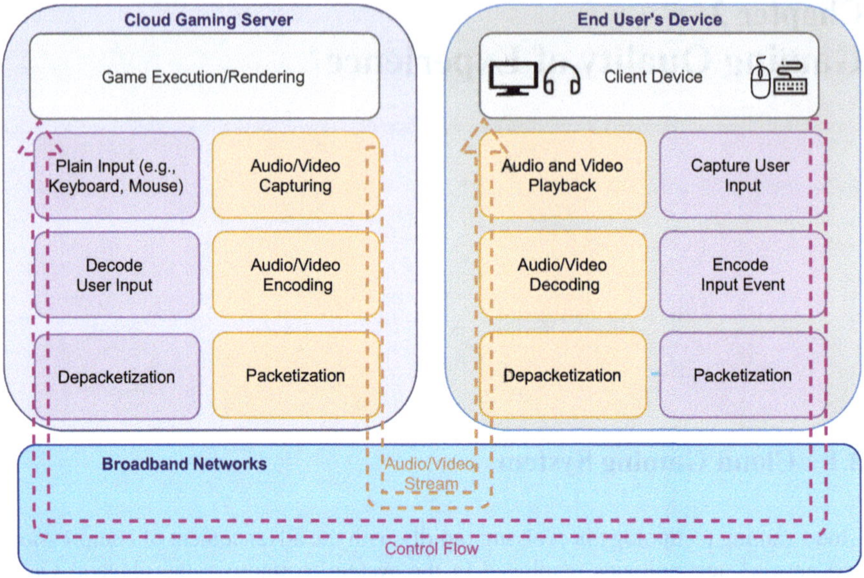

Fig. 2.1 Abstract framework of a cloud gaming system

latency, which might not be desirable for an interactive cloud gaming service. Thus, in cloud gaming services, packet loss typically results in frame losses. As a choice of network control protocol, the Real Time Streaming Protocol (RTSP) is considered in this book. However, this might change over the years with the advancement of web browser-based multimedia transmission technologies, as Google uses Web Real Time Communication (WebRTC) for its cloud gaming service, Stadia.

All processes between the client and the server introduce latency to the service. The delay typically is due to multiple processes between client and server such as packet transmission and propagation, video encoding/decoding, (de)packetization. Thus, an additional delay is introduced in the cloud gaming service compared to a traditional gaming system, e.g., game console. This additional delay could severely impact players' interaction with the system, which will be discussed in the following chapters.

2.2 Concept of Quality

The term "quality" is widely used in academia, while there is no general consent over the definition. Depending on the service, product, or circumstance, the definition of quality is adjusted and redefined in literature. A widely used definition of quality was introduced by Juran [4] as "fitness for use". In this definition, the term "quality" is associated with user requirements, and implicitly customer satisfaction is foreseen. However, this definition neglected the product and service cost, which led to a revision of the definition by Ishikawa and Lu as "fitness for use at an acceptable price" [5]. The definition of "quality" has changed over the years in different domains, while typically the customer is the central point of the definition.

The term "quality" in multimedia and telecommunication domains has a historical path from a technology-centric approach to a human-centric approach. In classical quality assessment in telecommunication systems, quality was related to the characteristics of a product (service, application, or device) to evaluate if the technical product fulfills the requirement that is expected without considering the satisfaction of users. This perspective led to a commonly agreed definition of quality given by the International Organization for Standardization (ISO) in 2005.

Quality
Degree to which a set of inherent characteristics fulfills requirements [6].

With the rise of new multimedia services and their diversity, more human-centric measurement of the quality is considered, leading to the adoption of newer concepts and methods for quality assessment of telecommunication services, which take into account the effect of human perception.

2.2.1 KPIs, QoS, and QoE

Nowadays, quality assessment has become a necessary step of not only Internet Service Providers (ISPs) but also service and application providers such as cloud gaming providers to ensure that their costumers are delighted with the provided service. In the early generation of quality assessment of telecommunication service, Key Performance Indicators (KPIs), such as delay, packet loss, and throughput, were measured by network and service providers. These KPIs are monitored continuously to ensure that the service or network provider commits to a certain level of quality for a given service which might be agreed upon in a Service Level Agreement (SLA). This technology-centric approach focuses on measuring the performance, and reliability of protocols and services that operate on top of the IP layer [7]. Thus, several efforts have been made to standardize these methods at the ITU-T, such as ITU-T Rec. Y.1540 and ITU-T Rec. Y.1541 define methods to assess the speed, accuracy, dependability, performance, and availability of IP packet transfer services

[8, 9]. These KPIs are the essential backbone of Quality of Service (QoS) models. QoS in the telecommunications domain is defined at ITU-R Rec. P800 [10] as:

Quality of Service
The totality of characteristics of a telecommunications service that bear on its ability to satisfy stated and implied needs of the user of the service.

QoS exclusively relies on the technical characteristics of telecommunication services for the user's satisfaction. The QoS models can help to create a better experience for users of a service, but these models fail to measure the quality from the user perspective. Therefore, a more human-centric approach to measure quality, termed as QoE is introduced to take in the perspective of human users perceiving and interacting with the multimedia systems and services. Quality of Experience is defined in a Whitepaper of the Qualinet group [2] as:

Quality of Experience
The degree of delight or annoyance of the user of an application or service. It results from the fulfillment of his or her expectations with respect to the utility and/or enjoyment of the application or service in the light of the user's personality and current state.

The QoE definition introduces new factors that are neglected in the QoS definition. QoE does not only take into account the technical factors but also the contextual factors as well as user related factors. In the classical QoS approach, user-related factors such as the customer's mood, personality, and environment are considered confounding and unwanted factors. Therefore, it was widely accepted that by averaging individual ratings into a MOS for a particular stimulus, the individual's influence would be eliminated, and it leads to a "typical" rating. The contextual factors are also controlled by strict requirements for the experimental environment, such as illumination, display size etc.

The new era of QoE led to efforts in developing quality prediction models that are much closer to human judgment than exclusively taking into account the technical characteristic of a system or service. This is now possible with a high-performance probing system that allows ISPs to monitor thousands of connections simultaneously and provide diagnostic analyses as well as QoE estimations [7]. However, peeking into all media streams transmitted over a network, capturing user-specific information touches the critical subject of customer privacy [7]. Therefore, many challenges still remain to develop a holistic QoE model for multimedia applications and services.

2.3 Taxonomy of Gaming Quality Aspects

With a general introduction to QoE in the previous section, QoE specific to the gaming domain is discussed in this section. QoE in gaming is different compared to traditional multimedia services, such as video streaming, as players are emotionally attached to their activities, and the players' actions significantly impact the Quality of Experience. Therefore, for a better understanding of the gaming QoE, a taxonomy of QoS and QoE aspects related to computer gaming has been proposed in [11], as illustrated in Fig. 2.2.

The taxonomy differentiates and categorizes influencing factors, interaction performance metrics, and quality features, in three layers. In any multimedia services, a series of influencing factors can be defined which impact the user's subjective QoE. The term Influencing Factor in the context of QoE has been defined in [2] as:

> **Influencing Factor**
> Any characteristic of a user, system, service, application, or context whose actual state or setting may have an influence on the Quality of Experience for the user.

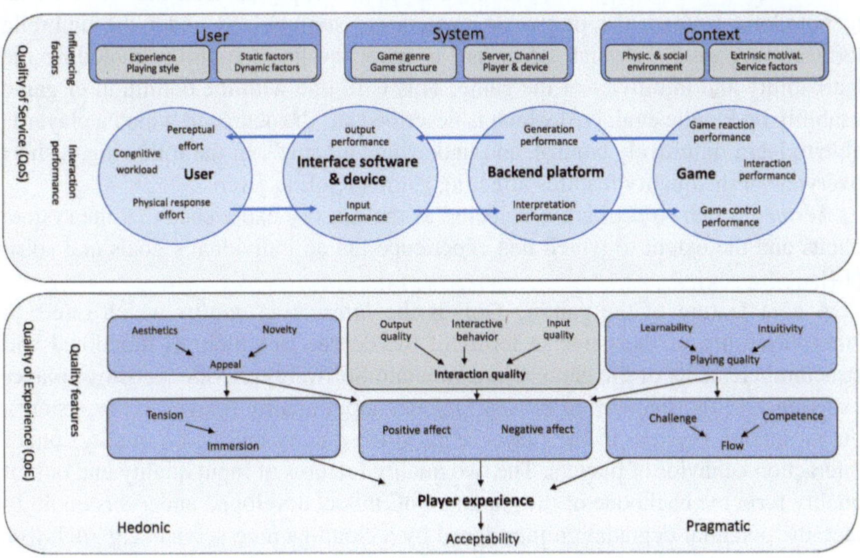

Fig. 2.2 Taxonomy of gaming QoE factors and aspects redesigned based on [11]

QoE as a multi-dimensional construct has many features specific to the service under scrutiny. The term QoE feature has been introduced in [2] as:

QoE Feature
A perceivable, recognized, and nameable characteristic of the individual's experience of a service that contributes to its quality.

The top layer of the taxonomy identifies the influencing factors categorized into three groups, user-specific factors, system-related factors, and context of usage. In the middle layer, the interaction between the user and the system, including any technical systems involved, are represented. The bottom layer includes the perceptual quality features of gaming QoE that may be affected by various influencing factors. Due to the importance of factors influencing gaming QoE modeling, they are listed and discussed in detail in Sect. 2.4. The relation of gaming quality features presented in the bottom layer is inspired by the Game Experience Questionnaire (GEQ) [12] as well as a range of quality features from the hedonic aspects, the joy of use, to pragmatic aspects such as ease of use.

Regarding the latter (pragmatic aspects) it must be noted that for a gaming application, the traditional view of usability is not suitable as there have to be some challenges and obstacles for a player to create a joyful experience. Thus, the concepts of efficiency and effectiveness do not apply here, as they would lead to boredom. Instead, the pragmatic aspects are summarized under the umbrella term *playing quality*, which takes into account the game usability including the learnability and intuitivity of the game. This is in line with the definition of game usability by Pinelle et al. [13] which is described as "the degree to which a player is able to learn, intuitively control, and understand a game". In the following, a short overview of the quality features affecting gaming QoE is given.

Aesthetics and Appeal are considered as the sensory experience that the system elicits and the extent to which this experience fits an individual's goals and spirit [14].

A core feature of the gaming QoE is the *Interaction quality* which refers to "the playability of the game in terms of the degree to which all functional and structural elements of the game (Hardware and Software) provide a positive player experience [14]". *Interaction quality* includes input quality (player to the system), output quality (system to the player, e.g., video quality and audio quality) under interaction behavior of players. The two quality features of input quality and output quality form the backbone of the gaming QoE model developed under this book. In fact, the potential degradation introduced by a cloud gaming service, e.g., delay or low video quality, would severely affect the interaction of players with the game, which points to the importance of these two quality features in the development of any gaming QoE model.

In the lower layer of quality features, the *player experience* is covered. Player experience is defined according to the QoE definition by ITU-T as "the degree of delight or annoyance perceived by the player after the gaming experience" and includes popular concepts such as *immersion* and *flow*.

There is still no agreement on a clear definition of the term *immersion*. Witmer et al. [15] define immersion as "a psychological state characterized by perceiving oneself to be enveloped by, included in, and interacting with an environment that provides a continuous stream of stimuli and experiences". In the gaming context, immersion often describes the degree of involvement with the game, which Brown and Cairns categorized into the three phases of engagement, engrossment, and total immersion [16]. The first phase of involvement requires the player to overcome the barrier of preference and learn how to play the game. To reach the engrossment state, the player needs to combine game features and master the control of the game in order to become emotionally attached. These two steps are necessary to reach the total immersion where the players are barely aware of their surroundings and themselves and emotionally attached to the game.

Flow is described by Hassenzahl as "a positive experience caused by an optimal balance of challenges and skills in a goal-oriented environment. In other words, flow is the positive user experience derived from fulfilling the need for competence (i.e., mastery); it is a particular experience stemming from the fulfillment of a particular be-goal." [17]. Flow depends strongly on the game design (e.g., difficulty adjustment) as well as user experience.

Fun as one of the main *positive affect* of playing games is typically the main goal of the gamers to play which could be associated with other positive affect such as delight, enjoyment, and so forth. Fun has been defined as "The positive feelings that occur before, during, and after a compelling flow experience" [18]. On the contrary, the *negative affect* such as frustration and boredom could negatively influence the experience of the player.

Finally, these aspects can affect the *acceptability* describing how readily a user uses a system or a service, which is influenced not only by player experience but also by other economic measures such as costs and service conditions.

For quality assessment of video games, it is required to select means to measure these features. Several measurement tools exist that can be used for the measurement of each feature, e.g., physiological measurements such as electroencephalography (EEG). The most widely used assessment tools are questionnaires. ITU-T Rec. P.809 [14] gives a long list of questionnaires that can be used in subjective tests to measure gaming quality features. Since it is not possible to cover all quality features affecting the gaming QoE with a single questionnaire, it typically depends on the aim of the study as to which questionnaire is selected and used in the experiment.

Since this book aims at the development of gaming QoE prediction models, as an essential step, it must be decided what features should be taken into consideration in the prediction of gaming QoE. Based on the gaming taxonomy, all presented features might be affected by the distortions in cloud gaming services. However, some of the features depend strongly on the game design, such as *aesthetics*, which is out of the scope of a cloud gaming QoE prediction model. In addition, some other features strongly rely on the information of the user that is not available to cloud gaming or network providers as two potential stakeholders of cloud gaming QoE prediction models.

As an example, because of the possible distortion due to end-to-end latency, limited bandwidth, and packet loss in cloud gaming services, the degree of involvement, consequently the immersion, could be negatively affected. However, it has to be noted that immersion (or state of involvement) requires knowledge about the player preference, state of the game, and experience with the game which is typically not accessible from the eyes of service and network providers. Similar to the immersion state, monitoring the flow state of players is not practically possible by network and service providers unless having access to the mastery level of participants, game state, and many other information that leads to a very complex and game/player dependent gaming QoE prediction model. This does not imply that these features have no impact on gaming QoE. However, the selection of features considered for the gaming QoE prediction model depends on several limitations such as availability of information to the stakeholders, the impact of each feature to overall QoE, and limitations in data collection and subjective testing. After the theoretical basis of the research is summarized in this chapter, the concrete decision made regarding the feature selection of the model is presented in Chap. 3.

2.4 Gaming QoE Influencing Factors

In the previous section, the taxonomy of gaming QoE was described that includes gaming QoE influencing factors as one of the main components influencing a player's perceived quality. In this section, the factors of the cloud gaming service that are influencing the gaming QoE are presented based on the ITU-T Rec. G.1032 [1]. This section is majorly written based on this Recommendation which the author of the present book contributed to between 2016 to 2017 [19, 20]. ITU-T Rec. G.1032 categorizes the gaming influencing factors into three groups of *human*, *system*, and *context* influencing factors that this section is structured accordingly.

2.4.1 *Human Factors*

A human influencing factor is "any variant or invariant property or characteristic of a human user. The characteristic can describe the demographic and socio-economic background, the physical and mental constitution, or the user's emotional state" [2]. In the following, some of the identified human influencing factors in the context of cloud gaming are summarized.

General Gaming Experience In the literature, a classification of "hardcore gamer" and "casual gamer" is widely used, distinguishing the classes based on the average playing time in a certain period. Despite the division's broad adoption, there exists no generally agreed threshold to delimit the two groups.

Experience with a Specific Game or Genre Each service is best evaluated by users who are familiar with how it should function. "Digital games" is a very broad term, and the complexity of the game under test can vary. For simple games, the knowledge of the player about the game under test may not matter because the gameplay is simple and easy to understand. On the other hand, in more complex games such as Massively Multiplayer Online Role-Playing games, it is intuitive that the users who are using it usually should be the ones who can best evaluate its quality, because some aspects of the system may not function properly and if the testing player does not know how it should function, he/she may not notice the degradation.

Intrinsic and Extrinsic Motivation Intrinsic and extrinsic motivation can have a strong influence on the QoE of gaming since the high diversity of games offers different kinds of fun and motivation to play them. For example, players who want to prove themselves by beating other players may not be satisfied when playing a purely artificial intelligence controlled game. It has to be noted that Extrinsic motivation is categorized as context factors in literature [21].

Static and Dynamic Human Factors Static human factors are static characteristics of a player such as demographic information (e.g., age, gender, and native language), while dynamic factors refer to user state factors (e.g., emotional status include boredom, distraction, curiosity).

Human Vision and Audition Visual perception varies depending on the characteristics of the visual stimulus. The sensitivity of a user to video/network artifacts differs between users. As an example, sensitivity to framerate as an encoding parameter depends on a user's Critical Flicker Fusion threshold [22]. Similarly, the human hearing capability varies in terms of pitch and loadness of the sound.

2.4.2 Context Factors

Contextual influence factors "are factors that embrace any situational property to describe the user's environment in terms of physical, temporal, social, economic, task and technical characteristics" [23, 24]. In the following, a summary of identified context influencing factors for cloud gaming is given.

Physical Environment Factors Physical factors refer to room characteristics (acoustics, and lighting) and usage situation (in-house, on the move, etc.).

Social Context The Social Context factors refer to the relationships to other players who are involved in the game, potential parallel activities of the player, privacy and security issues, which might be particularly relevant in multiplayer games [21]. Social context is one of the influential QoE factors that is difficult to measure objectively. Therefore, the influence of other factors (especially technical factors) on gaming QoE is hard to measure in the presence of social context. Hence, it is recommended to avoid this factor in gaming QoE studies, unless the main focus of the work is to study the influence of such factors [14].

Service Factors An influence of these factors (e.g., ease of access, availability, pricing) on customer satisfaction has been shown for online game services [25] and is likely present especially in the context of cloud gaming.

Novelty Novelty, which means that the user experience is improved when a new technology is introduced, has an impact on quality judgments, not because of any actual improvement in learning or achievement, but in response to an increased interest in the new technology [26]. Thus, it can be expected that the impact of quality factors on perceived quality will be different for any "new" technologies and services. An example of such a new technology is Virtual Reality (VR) gaming using HMDs.

2.4.3 System Factors

System influencing factors refer to "properties and characteristics that determine the technically produced quality of an application or service" [24]. The factors in this section contain the game, which is being played by the user, and also the whole setup and user-perceivable design of the hardware and software. Below, an overview of identified system factors is provided.

2.4.3.1 Game

Similar to other types of media, the content of a game (e.g., mechanics, dynamics, aesthetics [27]) decidedly influence a player's gaming experience. However, differing from other types of media, the content and the underlying technical implementation are strongly interwoven: even scenarios from the same game often employ entirely different stacks of software to generate the gaming experience [28]. In addition, several game design aspects have an impact on a game to be sensitive towards a certain type of degradation in cloud gaming services. A comprehensive discussion about the game characteristics that may have an impact on the player's experience in a cloud gaming service is given in Sect. 3.6.

Game Mechanics and Rules Djaouti et al. [29] introduced a rule-based game classification called "gameplay bricks" splitting the games into several fundamental elements such as moving and shooting. In total, ten gameplay bricks are classified into two categories: rules stating *goals*, including avoid, match, and destroy, and rules defining the *means to reach* the goals consisting of create, manage, move, random, select, shoot, and write. These largely influence and determine game outcomes and are individual to each game.

Game Genre Genre classification is a broad term and is not precise enough to characterize the games. However, in the absence of appropriate game classification, game genres can be used as a basic criterion of content selection in experimental design. It must be noted that several game interactions (game bricks) could be a part of a single game (genre) while the impact of technical parameters such as delay for each of these interactions most likely is not the same. Thus, the game genre alone will not be sufficient to characterize the sensitivity of the game towards technical parameters.

Temporal and Spatial Video Complexity Video complexity plays an important role in video streaming services, especially when bitrate and other encoding factors are considered. In contrast to traditional video content, gaming content has special characteristics such as an extremely high motion for some games, special motion patterns, synthetic content, and repetitive content, which makes the state-of-the-art video and image quality metrics perform weaker for this special computer-generated content [30, 31]. Video games are usually created based on a pool of limited objects that appear in different scenes of a game. Therefore, there is a high similarity of different scenes of a particular game with respect to the spatial domain. In addition, for many video games, due to the same design style across a certain game, the game shares similar visual features such as background scene, color diversity, and pattern of motion [30].

Temporal and Spatial Accuracy Temporal accuracy is defined as the time required to complete an action. Spatial accuracy is the degree of accuracy required to complete the interaction successfully [28]. These two factors might determine the playability of a certain game under delay.

Pace Pace is referring to the speed of gameplay. Although this factor seems to be similar to the two previously mentioned factors (the temporal accuracy concept as well as temporal complexity of games), there exists a small difference between pace and the two other factors. First, the pace has to be seen as a speed in one game type (or one game genre in general) which means that there could be two games with the same temporal complexity while their paces are not comparable (e.g., a racing game and shooting game with same temporal complexity). Also, a game with a high-required reaction time does not necessarily have a high pace (e.g., a static scene with some blinking points that have to be hit).

Visual Perspective of Player Based on the perspective of the camera, games have been classified into First-person Linear Perspective, Third-person Linear Perspective, and Third-person Isometric Perspective [32]. The perspectives of the game are very important in cloud gaming and an interplay with video coding can be assumed.

Aesthetics and Design Characteristics Aesthetics is the sensory experience the system elicits, and the extent to which this experience fits individual goals and spirit

[33]. Aesthetics and design characteristics describe the design of a game, which can be experienced by the player, and is commonly specified by design experts.

Learning Difficulty The required time to learn how to play a game is a critical criterion when aiming at a short interactive test. Games such as racing games do not necessarily need a long time to learn the game rules, actions, and game elements because of a limited number of control buttons and game rules. On the contrary, there exist games that need a long time to achieve the game goals and to learn the game elements. This factor should be considered in the design of subjective assessments.

2.4.3.2 Playing Device

Portability, Size and Input modalities The continued success of portable gaming devices suggests that the benefit of mobility outweighs the limitations of a portable device for a group of players. The success of smartphone gaming may also be attributed partly to the portability of the devices. The size of a handheld device has been shown to exert an influence on playing test participants' ratings [34]. Modalities used for game input differ vastly in terms of attributes like precision, speed, and feedback. In the past, new forms of input (e.g., gesture control) have repeatedly enabled new playing paradigms. Unless it is the object of investigation in a study, all three factors should therefore remain constant.

Output Modalities The availability of output modalities and their technical attributes confine the perceivable experiences. A study showed that participants wearing a VR headset perceive higher levels of immersion than players of the same simulation game using a conventional 2D screen [35].

Display For video quality assessment, it was shown [36] that the perceived quality is strongly influenced by the viewing distance, display size, brightness, contrast, sharpness, screen resolution, refresh rate, and color. Not only the size of the display but also the refresh rate of the display can bring higher quality if the framerate is high.

2.4.3.3 Network Transmission

Given the highly interactive nature of cloud-based games, a network distortion can negatively influence the user experience. Four main distortions that need to be taken into account are delay, jitter, bandwidth, and packet loss.

Delay The delay perceived by an end-user corresponds to the delay from user commands' execution to the visible game event shown on display. While most studies focus on the impact of networking delays, often additional system components contributing to overall end-to-end delay are neglected due to difficulty in measurement (e.g., system tick rate, processing, and rendering delay). The influence of delay on the QoE strongly depends on the game characteristics, as discussed earlier. Therefore, investigating delay as an influencing factor of QoE without considering the characteristics of the game is not valid, and results are not comparable with other studies. In addition to game characteristics, the perceived delay may differ considerably depending on the usage of input devices for playing.

Jitter Jitter has a perceivable influence on online gaming and cloud gaming experience [37, 38]. Depending on the client implementation, jitter may also result in a less smooth visual appearance of the game as frames are displayed at varying intervals.

Bandwidth The impact of bandwidth restrictions on gaming QoE has been proven in numerous studies [39, 40] and has been shown to be a contributing influence factor to gaming QoE. Depending on the employed mechanism to deal with limited bandwidth, it could result in packet loss, delay (due to buffering mechanism), and video artifact due to video compression.

Packet Loss Network packet loss has a significant impact on gaming QoE while playing intensive games such as FPS, with packet loss smaller than 1% causing serious degradation of user experience [41]. Increased packet loss causes severe degradation of graphics quality of the games, culminating with a lower framerate and an unplayable gaming experience.

2.4.3.4 Compression

Given bandwidth limitations, different QoE-driven codec configuration strategies may be specified for various types of games. There are differences in terms of performance of different codecs for gaming streaming application; some discussion can be found in [42]. In this section, a few important encoding parameters that need to be considered for cloud gaming quality assessment are presented.

Framerate The framerate has a significant impact on a user's performance and consequently influences the QoE. However, perceiving the difference between very high framerates strongly depends on human vision abilities, a player's commonly used gaming setup, and game characteristics (especially the game temporal complexity).

Encoding Resolution The encoding resolution is an important parameter, which affects the video quality in all streaming applications while it has little impact on user performance [43]. It should be noted that display specifications such as refresh rate and display resolution would be substantial when investigating the impact of framerate and resolution on the gaming QoE.

Rate Controller Modes There exist some strategies to control the video streaming rates in order to reach a certain target quality for a given bandwidth. Three different rate controller modes were implemented in the software implementation of H.264/MPEG-4 AVC, x264, including Constant Quantization Parameter (CQP),

Constant Rate Factor (CRF), and Constant Bitrate (CBR); all of them result in a different quality level. It has to be noted that Variable Bitrate (VBR) is excluded since it is out of the scope of cloud gaming services. CQP is a straightforward strategy that encodes the video based on constant Quantization Parameters (QP) and leads to variable bitrate and quality. CRF attempts to keep the quality at a certain level by adjusting QP in a video sequence. Finally, CBR tries to encode the video by keeping the bitrate constant. Rate control strategies do not only affect the QoE, but they also have an impact on the overall delay of a gaming service. For real-time application, NVIDIA introduced a hardware implementation of H.264/MPEG-4 AVC using GPU accelerator engines, named NVENC, which offers CBR in a low latency high quality (llhq) preset controller mode for fast encoding. The llhq preset uses multiple passes for encoding, which leads to more efficient bit allocation while using accelerator engines for real-time encoding.

GoP Group of Picture (GoP) is a structure that specifies the order of inter and intra frames (I, B, and P frames) in a video sequence. The llhq preset that is discussed earlier uses no Bidirectional-Frames (B-frames) to avoid introducing latency and uses an infinite GoP for real-time coding.

With the introduction to QoE influencing factors in the context of cloud gaming, in the next section, the design of subjective experiments for assessing cloud gaming services is presented.

2.5 Subjective Assessment

Subjective tests assessing multimedia services are typically conducted in a controlled environment, i.e. a neutral laboratory setting, using pre-defined game scenarios and test participants. For gaming quality assessment, ITU-T Rec. P.809 provides information that needs to be considered when conducting an experiment for gaming QoE assessment. In this section, a short overview of the ITU-T Rec. P.809 is given. ITU-T Rec. P.809 defines two test paradigms to assess gaming quality:

- *Passive viewing-and-listening tests* with pre-defined audio-visual stimuli passively observed by a participant.
- *Interactive tests* with game scenarios interactively played by a participant.

The passive paradigm can be used for gaming quality assessment when the impairment does not influence the interaction of players with the game as well as when the study targets the passive video streaming services such Twitch.tv. For example, the output quality feature (cf. Sect. 2.3), when only spatial video quality is triggered, can be assessed using the passive viewing-and-listening test which allows the researchers to collect more subjective data due to the short stimulus (cf. Sect. 2.5.2). In addition, the experimenter would have more control over the scene content as the same content with the same temporal and spatial video complexity will be shown to all participants. The interactive test must be used when other quality features are under investigation, such as input quality, playing quality, immersion, and flow.

2.5.1 Experimental Setup

In general, the assessment methods of ITU-T Rec. P.910 [3] and ITU-T Rec. P.911 [44] are recommended to be followed to set up the experiment environment. Thus, a neutral environment, e.g., sound-shielded rooms with daylight imitation, is recommended except in the cases that a real-life gaming situation is the purpose of the test. As a choice of the display, a 24 inches monitor and a viewing distance of three-times the video height are recommended.

Participants are required to be screened and selected based on their game experience, and dependent on the purpose of the test, the target group should be selected. Prior to the test, participants are required to be screened for normal visual acuity and normal color vision. In case audible stimuli are presented, pre-screening procedures such as audiometric tests need to be considered.

The quality perception of stimuli is influenced strongly by their content due to several types of games with differences in design, game mechanics, and game characteristics as well as the diversity of players' preferences. The selection of games and scenario of a game in interactive tests requires very careful consideration to meet the following criteria:

- A scenario of a game should be chosen in which participants are able to experience the game in similar ways repeatedly. Thus, the participant should not have too much freedom, causing the scenario to be too difficult for one while too easy for another participant just because of their gameplay decisions.
- When encoding parameters are under investigation, it is recommended to select a scene of a game that is representative in terms of video complexity. For example, in an action game, a scene in which the player is only looking at a map is not recommended to be selected.
- It is recommended to avoid horror games or overly violate games, not only for ethical reasons but especially when physiological measurements are used to capture the user's state.

ITU-T Rec. P.809 recommends to use experimental designs for the randomization of the stimuli, such as complete randomized design, Latin, Graeco-Latin, and Youden square designs, replicated block designs.

2.5.2 *Passive Viewing-and-Listening Tests*

ITU-T Rec. P.809 recommends a passive viewing-and-listening paradigm for assessing the output quality to ensure that every participant is rating exactly the same content in an experiment that is easy to conduct on a large scale. In addition, the playing abilities of the participants will not influence the outcome of the study. In case that video compression is under investigation, it is recommended to select scenarios of the games that cover an appropriate range of spatial and temporal video complexity. Spatial and temporal indexes are provided in ITU-T Rec. P.910 [3] that can be used for screening the video complexity.

The stimulus duration, as denoted by ITU-T Rec. P.809, depends on the test purpose and game content. It is recommended to select the stimulus duration that covers different game mechanics that would be visible in an interactive scenario. Thus, a duration of 10 s which is typically used in video quality tests [3] may not be enough to meet this requirement. Schmidt et al. [45] as well as Claypool [46] have shown that there is a significant difference in video quality ratings for different stimulus durations. Thus, ITU-T Rec. P.809 recommends a minimum of 30 s stimulus duration for passive tests to ensure that the recorded scene represents the game in terms of video complexity while allowing participants to detect degradations similarly as during interactive tests.

For passive viewing-and-listening tests, it is recommended to give instruction about the game rules and game objectives to allow participants to have similar knowledge of the game. In addition, this will prevent the participants from asking about the game during the test session. One of the very common mistakes of participants in video quality assessment of gaming content is understanding the difference between video quality and graphic quality (e.g., graphical details such as abstract and realistic graphics). Participants should be instructed to not rate the graphic quality which relates to the game design but rather the video quality, which may be degraded due to compression artifacts such as blockiness. This is especially important when an abstract game (e.g., the game Cuphead) or a game with a blocky design (e.g., the game Minecraft) is used in the experiment. Finally, the participants need to know the general information about the experiment, such as details about the questionnaire and test structure. To summarize, the following information should be provided prior to the passive test:

- Information about the experimental details such as assessment methods (e.g., questionnaire) and session duration.
- Information about the objective of the games, e.g., enemies and obstacles. It is also recommended to provide it in a written text or recorded video to ensure that the same information is given to all participants.
- The experimenter should not suggest that there is a correct rating or provide any feedback as to the "correctness" of any response.

2.5.3 Interactive Tests

In the interactive paradigm, participants play a game scenario within a limited period of time, which varies depending on the purpose of the test and rate their experience a posteriori. ITU-T Rec. P.809 defines two types of long and short interactive tests, which can be chosen depending on the aim of the subjective experiment.

- *Short interactive* is recommended for assessment of interaction quality (e.g., the impact of delay on the control). ITU-T Rec. P.809 suggests a duration of 90 to 120 s.
- *Long interactive* is recommended when aiming for assessment of all QoE features described in Sect. 2.3 such as immersion and flow. The recommendation states that the duration of a stimulus depends strongly on the game and player experience. However, 10 to 15 min could be considered a reasonable stimulus duration.

One common problem in interactive experiments is the learning effect, which happens when a player is new to the game and learns how to play the game during the experiment. In addition, due to differences in participants' gaming experience level with the test games (or the genre of the test games), their experience could vary strongly. Therefore, before starting the subjective experiment, to ensure that the participants have similar knowledge of the game and reduce learning effects, the rules and control of the game should be given to the participants. Moreover, participants should know about the experimental details before the start of the experiment. Therefore, the following information should be provided prior to the test:

- Information about the experimental details such as assessment methods (e.g., questionnaire), and session duration.
- Information about how to control the game via the input device(s), describe the games' objective, e.g., enemies and obstacles. It is also recommended to provide a short training session.
- The experimenter should not suggest that there is a correct rating or provide any feedback as to the "correctness" of any response.

2.5.4 Interactive vs Passive Paradigm

ITU-T Rec. P.809 suggests using passive viewing-and-listening tests instead of interactive tests when only spatial video quality (as part of output quality feature) is under investigation. However, it raises the question if the participants rate the video quality of an impaired video signal similarly when they play a game compared to the case that they watch the gameplay.

Schmidt et al. [45] conducted a study that compares the result of video quality for a passive viewing-and-listening test and an interactive test. In this study, two games

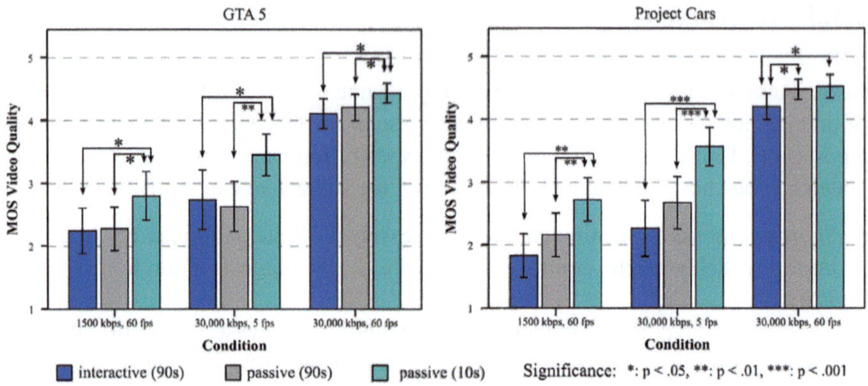

Fig. 2.3 Bar plots for Mean Opinion Scores of video quality and 95% confidence interval (labels (1–5): bad, poor, fair, good, excellent), taken from [45]

were selected, GTA 5 and Project Cars, under a controlled scenario that the players are limited to a concrete and repetitive task to ensure that a very similar audiovisual content will be experienced in both interactive and passive tests. The passive test was conducted with two stimulus durations of 10 s and 90 s, while the interactive test was conducted only with 90 s stimulus duration. Figure 2.3 illustrates the bar plots of video quality ratings for the three tested conditions. Based on the barplot, it can be concluded that the ratings of video quality during the interactive test and the long passive test (90 s) are very similar. Post-hoc tests showed no significant differences for video quality ratings between long passive and interactive tests, $p > .05$, except for the high bitrate condition of Project Cars. It has to be noted that "Project Cars" is a car simulation racing game in which the car in the game has slight inertia towards a change of direction compared to most of the popular racing games in the market. This was also mentioned in a short interview conducted after the experiment by participants that the inertia of the car in the "Project Cars" might negatively influence their ratings. This could potentially explain the slightly lower ratings of participants in the interactive test compared to the passive test for the game, "Project Cars."

While the study was conducted with a limited number of games and tested stimuli, the result confirms that if the stimulus duration of passive test is long enough, the video quality would not significantly differ between interactive and passive tests. It has to be noted that the interactive study was conducted under a very strict scenario plan, in which participants did not have too much freedom, which could potentially have resulted in a diverse gameplay scene in terms of video complexity. The video quality in an interactive test could be significantly influenced by the change of video complexity based on the gameplay of participants, which was avoided in this study by the strict scenario provided to participants.

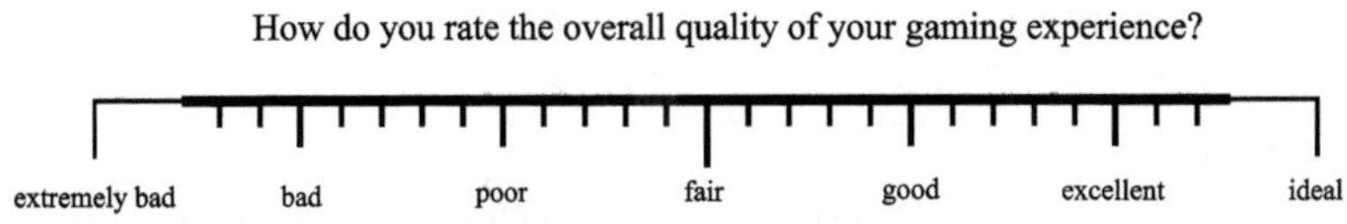

Fig. 2.4 The scale and question to assess the overall gaming quality following the EC scale [14]

2.5.5 Scaling for Gaming QoE Assessment

The application of summative subjective methods like the determination of the "overall quality" of audio-visual content has been used and standardized for quality assessment of speech and audio-visual material. These methods allow test participants to express their opinions about the quality aspects of multimedia stimuli by rating one or a limited number of perceptual aspects on predetermined scales. One of the widely used scale in quality assessment of speech, video, or audio-visual services is a five-point ACR scale ranging from "bad" to "excellent". The collected judgments of a certain stimulus on the ACR scale are then combined to the arithmetic mean, which is commonly referred to as the MOS [47]. ITU-T Rec P.809 recommends a 7-point Extended Continuous scale (EC scale) instead of an ACR scale for quality assessment of gaming applications. The scale for assessing the overall gaming QoE is shown in Fig. 2.4.

In ITU-T Rec P.809, a list of questionnaires that can be considered for quality assessment of gaming application is compiled. While for a specific quality feature such as immersion, dedicated questionnaires are available, to cover most quality features of the gaming QoE taxonomy (cf. Sect. 2.3), the In-game GEQ (iGEQ) [12] could be considered to be used.

In addition, Schmidt [48] developed a questionnaire named Gaming Input Quality Scale (GIPS), measuring the input quality during the development of ITU-T Rec. P.809, which turned out to be a valuable instrument to quantify the impact of network-related influence factors on players' experience. Moreover, ITU-T Rec. P.918 [49], provides a questionnaire and a methodology to measure the sub-dimensions of the video quality. These two questionnaires are described in the following sections.

2.5.5.1 Gaming Input Quality Scale (GIPS)

Exploratory studies have shown that the participants strongly value the interaction quality, also referred to as playability, for their judgment of overall gaming QoE. Interaction quality includes input quality (player to the system), output quality (system to the player) under the interaction behavior of players. Contrary to output quality, which can be assessed using questionnaires developed to assess the different dimensions of audio or video quality, e.g., ITU-T Rec. P.918 [49], there has been no validated questionnaire available to assess the input quality in the literature until

Table 2.1 List of GIPS items [48]

Code	Label
Controllability	
CN1	I felt that I had control over my interaction with the system
CN2	I felt a sense of control over the game interface and input devices
CN3	I felt in control of my game actions
Responsiveness	
RE1	I noticed delay between my actions and the outcomes
RE2	The responsiveness of my inputs was as I expected
RE3	My inputs were applied smoothly
Immediate feedback	
IF1	I received immediate feedback on my actions
IF2	I was notified about my actions immediately

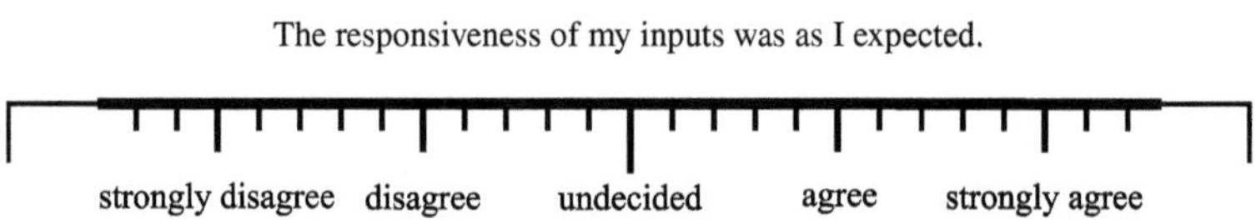

Fig. 2.5 Example of item and scale used for the GIPS [48]

recent years. A questionnaire assessing the input quality was developed by Schmidt in [48] using the sub-aspects of controllability, responsiveness, and immediate feedback. Responsiveness and immediate feedback describe the temporal aspects of the feedback a player receives after performing an action, e.g., a mouse click or a keystroke. The response of the game (system) should be available immediately after the player performs an action (input event) [48]. In other words, responsiveness can be defined as reacting quickly in the way that is needed and suitable for a gameplay situation, while immediate feedback describes the instantaneous reaction of the game that a player receives after performing an action.

The perceived controllability is the degree to which a player is able to control a game using the given input device and available interaction possibilities. It describes whether the performed input action resulted in the desired outcome. The controllability does not relate to the learnability of the controls nor to autonomy (freedom or power over something).

Table 2.1 contains the list of the items of the GIPS questionnaire. The rating scale is based on an agreement Likert scale using five labels ranging from strongly disagree to strongly agree on a 7-point continuous scale, known as extended continuous ACR (EC-ACR) scale. An example of the scale is provided in Fig. 2.5.

Table 2.2 Perceptual video quality dimensions introduced in ITU-T Rec. P.918 [49]

Name	Description	Example impairment
Fragmentation	Fallen apart, torn and blockiness	Low coding bitrate
Unclearness	Unclear and blurry image	Upscaling effect using bicubic function
Discontinuity	Interruptions in the flow of the video	Low framerate
Noisiness	Random change in brightness and colour	Circuit noise
Suboptimal luminosity	Too high or low brightness	Over- and under-exposure

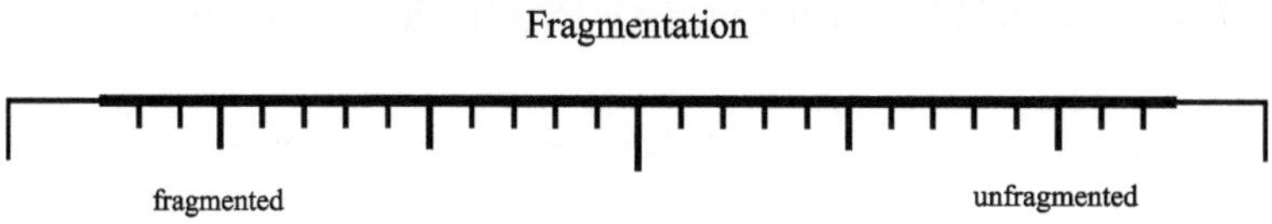

Fig. 2.6 Item and scale for Video Fragmentation according to ITU-T Rec. P.918 [49]

2.5.5.2 Dimension-Based Video Quality Evaluation

The output quality could be assessed through viewing-and-listening (or audiovisual quality) tests as discussed earlier. To get detailed insights about the influence of encoding parameters on the perceived video quality, ITU-T Rec. P.918 proposes a dimension-based subjective quality evaluation for video content, describing the identification of relevant perceptual video quality dimensions. The dimension-based subjective quality evaluation method (DBSQE-V) was developed using a pairwise similarity experiment with a subsequent multidimensional scaling (MDS) and a semantic differential experiment followed by a principal component analysis (PCA). Table 2.2 summarizes the identified dimensions, which later can be assessed using a direct scaling method.

Each dimension is explained to the participant in an introduction in written form using describing adjectives and in form of example videos. The rating scales are based on bipolar scales, using the dimension name as an item label and antonym pairs to describe the range of the scales. An example of the scale for the video fragmentation dimension is given in Fig. 2.6. For more information about the method, the reader is referred to ITU Rec. P.918 [49], where insights about the usage of the different rating scales as well as the test procedure are given. The dimension video discontinuity is considered as temporal video quality, whereas the remaining dimensions form the spatial video quality. Video noise (e.g. pepper and salt noise or flickering) is not a common artifact in subjective tests of cloud gaming services and luminosity is a game design aspect and is not seen as an artifact. Therefore, for cloud gaming quality assessment these two items are removed in the conducted studies using the DBSQE-V approach [50, 51] to avoid fatigue due to a lengthy questionnaire.

Video fragmentation is defined by synonyms such as fallen apart, torn, and blocky which could be triggered by a low bitrate or slicing effect of packet loss (cf. Sect. 5.1). Video unclearness is defined by synonyms such as unclear, blur, and smeared images which can be triggered by image or video upscaling methods such as bilinear interpolation. Video discontinuity is defined by synonyms such as jerky, stuttering, and wobbly, which can be triggered due to low framerate or frame loss due to packet loss.

2.6 Overview of Gaming Quality Prediction Models

In the past few years, several attempts have been made to develop quality models that predict gaming QoE. In this section, a short overview of these models is given.

While several quality features construct the gaming QoE as discussed in Sect. 2.3, two important quality features for gaming QoE are considered in the recently published standardized quality model for cloud gaming services, ITU-T Rec. G.1072 [51], input quality and output quality (only video quality is considered as output quality in G.1072). Input quality takes into account mostly the network influencing factors that may affect the interaction of players with the game. In addition to the input quality, the output quality feature is important which refers to the audio-visual aspect of the video stream. It has to be noted that gaming QoE is a multidimensional construct consisting of several dimensions such as immersion, flow, and presence. Some of these dimensions are mainly influenced by the game content itself, e.g., due to the game design, challenges, and rules, or by preferences of the player. However, from the perspective of a cloud gaming provider, who is not per se a content creator, these dimensions are not feasible to be measured or tracked. Therefore, most of the state-of-the-art (SoA) researches model the gaming QoE based on the input and output quality that form the interaction quality.

The prediction models for gaming QoE in literature can be categorized into three different groups, depending on the targeted quality feature. While some models target only the relevant parameters that affect input quality, e.g., by delay and packet loss, other models predict the gaming video (or audiovisual) quality based on information such as network/compression parameters and signal information. Finally, there is a limited number of models that consider both output and input quality.

The first group is mostly designed to measure the impact of delay and packet loss on user experience in cloud gaming and online gaming applications. One of the pioneer works on modeling the user interaction with the computer is Fitts' law [52], which describes the time to select a target as a function of the distance and the size of the target item. Later, Claypool [53] used the concept and introduced two characteristics called precision and deadline to define the sensitivity of a game to delay. Deadline is defined as the time required to complete an action that is the length of time it takes to achieve the final outcome of the action. Precision is the degree of accuracy required to complete the interaction successfully. More recently, Claypool [53, 54] modeled the influence of delay on user performance by considering velocity and angle in addition to the distance and size from Fitts' law. Claypool et al. [54] attempted a more fundamental approach to develop a model that can predict the user experience of a task that requires selecting a moving target with a mouse under different levels of delay. Selecting a moving object is a fundamental action for many computer-based multimedia applications such as computer games. The paper presents a user study that investigates the effects of delay and target's speed on the time to select the target and develops a model of target selection time with exponential relationships based on two parameters of delay and target speed. Claypool [53] complements the research by extending the study and using a game controller with a thumbstick instead of a mouse as an input device. The results revealed a very similar trend of exponential relation of target selecting time with the increase of delay and target speed. Comparing the results of the two papers, Claypool concludes that "the model's relationship between selection time, delay, and target speed holds more broadly, providing a foundation for a potential law explaining moving target selection with delay encountered in cloud-hosted games".

Another study was conducted by Long and Gutwin [55], where they modeled the influence of latency on gamers' performance by using the game speed. However, the work was limited to a constant game speed. Schmidt et al. [28] built a model based on three quality features for three different games under different levels of delay. Although the model only took into account delay as an influencing factor, the results showed that not only games have different levels of sensitivity towards delay, but also within the same game, the delay sensitivity might vary for different game levels. In fact, changing the pace within the same game can lead to stronger differences in respect to the delay sensitivity than using another game type.

In addition, Claypool et al. [56] conducted two studies to measure the impact of latency on a commercial cloud gaming service, OnLive,[1] as well as an academic cloud gaming service, GamingAnywhere [57]. The results revealed that latency linearly affects the user experience and gaming QoE for two third-person avatar games.

The second group of models targets the effect of video compression on output quality or overall gaming QoE. The transmission bandwidth is a major factor in cloud gaming, where the entire visual and audible output of a game is streamed over the network to the player. In this situation, the video compression needs to be adapted to the available end-to-end network bandwidth. The lossy compression causes degradations on the audiovisual scene. The early models in this group concentrate on the development of models that rely only on the encoding parameters, while the recent works do not only use encoding parameters for quality prediction but also the signal and bitstream information.

Slivar et al. [58] developed a gaming QoE prediction model that takes into account multiple parameters, including bitrate, framerate, game type, and player

[1] OnLive was one of the most the popular cloud gaming service between 2012 to 2015 when it was shut down and most of the assets were sold to Sony Computer Entertainment.

skill. The model was developed based on a large subjective laboratory study involving 52 players. Based on the data analysis of the subjective test, Slivar et al. [58] concluded that the game type is an important influencing factor that should be taken into account when evaluating the QoE of cloud games. In addition, the result revealed that there is no linear relationship between framerate and QoE and this relation depends on video bitrate. This result is in line with the finding in a subjective test conducted by Zadtootaghaj et al. [59]. Zadtootaghaj et al. [59] conducted a subjective experiment in a laboratory environment investigating the framerate and bitrate on gaming QoE. A QoE model is developed based on the quality features of video quality, positive affect, and game responsiveness. Then, each quality feature was predicted based on encoding parameters and signal information.

With respect to video quality models of gaming streaming services, as discussed earlier, recently, there has been a shift from parametric-based models, that only take the compression/network parameters into account, to signal-based models in order to predict gaming video quality. This shift is mostly due to the rising popularity of passive gaming video streaming services (e.g., E-sport), thus requiring accurate video quality metrics designed for gaming content. Barman et al. [31, 60, 61] evaluated multiple well-known signal-based image/video quality models on a gaming video quality dataset. The results revealed a medium to high performance of Full-Reference (FR) quality models, while No-Reference (NR) models perform poorly on gaming quality datasets. A short overview of the two research works that Barman conducted in collaboration with the author of this book in 2018 is described in Sect. 2.7.

Göring et al. [62] proposed an NR metric named $Nofu$, which is a pixel-based video quality model for gaming content. $Nofu$ uses 12 different frame-based values and a center crop approach for the fast computation of frame-level features. It further uses frame-level features pooling at video-level and feeds the features to a machine learning-based model for quality prediction. $Nofu$ takes into account the gaming-specific features by handcrafting low-level features that are common for gaming content, such as static areas (e.g., heads-up display in games). Barman et al. [63] proposed two NR signal-based video quality models for gaming content, $NR-GVSQI$, and $NR-GVSQE$. The two proposed models are designed using supervised learning algorithms based on the values of MOS and a video quality metric named Video Multimethod Assessment Fusion (VMAF). $NR-GVSQE$ is trained based on frame-level features as input and VMAF values as a target, whereas $NR-GVSQI$ is trained based on MOS values. Damme [64] proposed a Reduced-Reference (RR) model that predicts the quality of streaming game videos based on a low-complexity psychometric curve-fitting approach. This is the first RR model in the literature proposed for gaming video streaming applications.

One of the early works that take into account both output quality and input quality is the parametric-based quality model named Mobile Gaming User Experience (MGUE) [65]. MGUE uses system parameters to predict MOS using three different video games in mobile cloud gaming scenarios. The model's input parameters can be grouped into four categories: source video factors, cloud server factors, wireless network factors, and client factors. The model follows the structure of the core equation of ITU-T G.107, known as the E-model (Equation 7–1 in [66]), and proposes a set of impairment factors, which degrade the experience of players. To accommodate the varying susceptibility of different game genres towards degradations such as latency or packet loss, the authors introduced a set of tuning variables for the three specific games in their experiment. As a consequence, the model is limited to the three games tested in their study, without proposing a general prediction model. Furthermore, the model has been criticized for unmotivated and undescribed fitting factors in the equations and has not been developed using actual mobile games but streamed PC games, which are designed for different display sizes, input devices, and different usage contexts. The structure of the model is also not built around perceptual dimensions; instead, it uses direct links between technical parameters and overall quality. These limitations of the MGUE model permit only a restricted use (i.e. for the games and use-cases it was developed for) and unfortunately support no conclusions about the perceptual influences caused by the involved technical parameters.

Arguably the most comprehensive work has been done in ITU-T Study Group 12, resulting in an opinion model that predicts gaming QoE based on several influencing factors ranging from encoding parameters and game types to network degradations. The result is published as a recommendation named ITU-T Rec. G.1072 [51], and will be further discussed in the following sections. Table 2.3 gives an overview of studies that aim at the development of quality prediction models for cloud gaming or video streaming services.

2.7 Gaming Video Quality Model

In the previous section, the efforts to develop a video quality model for gaming content was discussed. Before the development of any new video quality model for gaming content, it is necessary to investigate how the SoA image/video quality models perform on gaming video datasets. This section investigates the performance of well-known signal-based image/video quality metrics on the gaming video quality dataset. First, a short overview of different types of video quality models is presented. Next, a comparative study of non-gaming and gaming videos using objective and subjective measurements is presented. Finally, the result and outcome of a comprehensive study evaluating multiple image/video quality metrics on a gaming video quality dataset is briefly discussed.

2.7.1 Classification of Video Quality Models

The objective quality prediction models can be classified into multiple groups according to the level of information that is available. Figure 2.7 illustrates the classification of objective quality metrics according to different levels of access to

Table 2.3 Overview of studies addressing (cloud) gaming quality prediction models following the structure overview provided in [58]

Author	Tested influencing factors		Scaling method	Description of model/Findings
	Network and encoding	User and context		
Lee et al. [67]	Latency	Game genre	fEMG	Model to predict real-time strictness of a game based on user input rate and game dynamics
Wang et al. [65]	Latency, packet loss framerate, resolution	Game genre	GMOS (Game Mean Opinion score)	Proposed a model for mobile cloud gaming user experience based on manipulated factors in the study
Claypool et al. [56]	Latency	Game genre, user device	7-pt. ACR (OnLive); and 5-pt. ACR (GamingAny)	Cloud-based games (regardless of the genre) are as sensitive to latency as highly sensitive shooting games in traditional online gaming
Slivar et al. [68]	Latency, packet loss	Player skill	5-pt. ACR	Modelled QoE as a linear function of network delay and packet loss
Hong et al. [69]	Framerate, bitrate	Game genre	7-pt. ACR	Proposed gaming quality model as a quadratic function of two video encoding parameters (framerate, video bitrate)
Slivar et al. [70]	Framerate, bitrate	Game type, player skill	5-pt. ACR	Modelled QoE as a linear function of video framerate and bitrate
Slivar et al. [58]	Framerate, bitrate	Game type, player skill	5-pt. ACR	Proposed a gaming quality model including graphics quality and fluidity of gameplay
Schmidt et al. [28]	Delay	Game type	7-pt EC-ACR (ITU-T P.809)	QoE model based on perception of delay, perceived difficulty, control
Claypool et al. [53, 54]	Delay	Game characteristics	5-pt ACR	Proposed a model to estimate target selection time based on responsiveness feature
Long et al. [55]	Delay	Game speed	5-pt ACR	Proposed a model to predict the time to react of participants
Zadtootaghaj et al. [59]	Resolution, framerate, bitrate	–	7-pt EC-ACR (ITU-T P.809)	Gaming QoE model, based on the positive affect, reactiveness and coding quality

Barman et al. [63]	Resolution, bitrate	Game scene complexity	5-pt, ACR, video quality	No-reference signal-based video quality model
Göring et al. [62]	Resolution, bitrate	Game characteristics (e.g. static areas)	5-pt, ACR, video quality	Light weighted no-reference signal-based video quality model
Damme et al. [64]	Resolution, bitrate	Game characteristics (e.g. static areas)	5-pt, ACR, video quality	RR signal-based video quality model based on curve-fitting

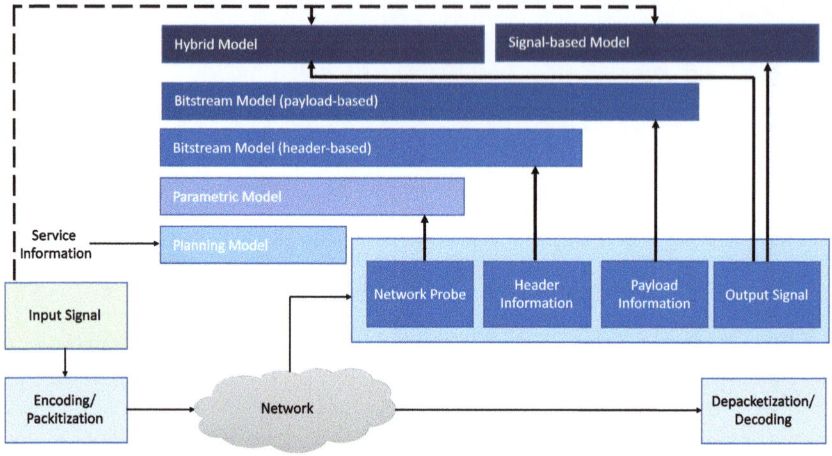

Fig. 2.7 Classification of objective quality assessment models according to different levels of information extracted from the media stream according to [71]

information. As the first group, planning models are meant to predict the quality of video streaming services based on the assumption of the network, client, and encoding parameters. These models are typically used by network planners to estimate the quality of the targeted service before establishing the service. The second category is parametric-based models that are similar to planning models but can be used while the service is running and by extraction of the network, client, and encoding parameters using network probes that are located in the network. If the bitstream information is accessible, typically, bitstream-based quality models are employed for quality assessment. Bitstream-based models are meant to be used for monitoring the quality of video streaming services. Bitstream information can be extracted from the payload or from the transmitted packet header. Finally, signal-based models are video quality models that process the video signal for the prediction of video quality. Depending on the knowledge about the reference signal, they can be categorized into NR models, RR models, and FR models. NR metrics require no knowledge from the original signal but only distorted/compressed signal. RR methods use some extracted features of the source signal and require full access to the encoded output signal to predict the video/image quality. FR methods are used when there is full access to both original and degraded signals, which typically have higher performance compared to RR and NR models. Figure 2.8 illustrates the classification of signal-based quality prediction models according to information available from the source. In addition to signal based models, hybrid models take advantage of the bitstream-based and signal-based information for video quality prediction.

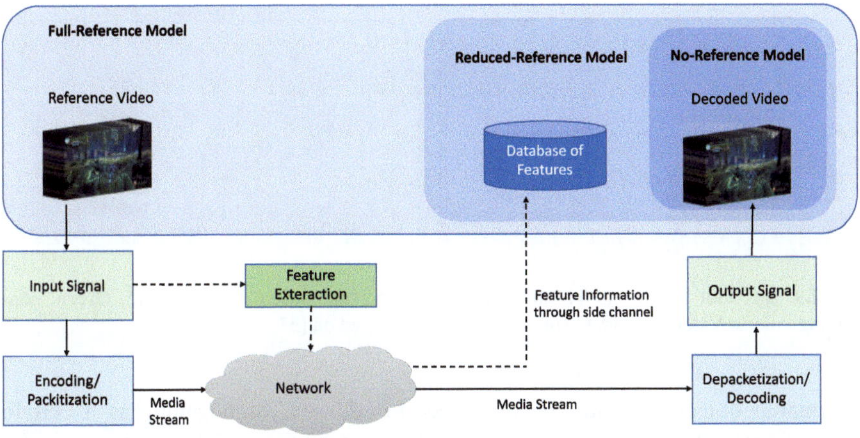

Fig. 2.8 Classification of the signal-based quality assessment techniques according to information available from the source

2.7.2 Comparative Quality Assessment of Gaming and Non-Gaming Videos

In 2018, Barman et al. [31] in a collaboration with the author of the present book conducted a study to compare the gaming content and non-gaming content for a video quality assessment task. To conduct such a comparison study, 30 pristine video sequences at 1080p resolution and framerate of 30 fps were recorded or collected from different datasets. As the first part of the analysis, the spatial information index (SI) and temporal information index (TI) for all pristine videos were measured. In the measurement of SI and TI values, only the frame-level prediction was calculated which is slightly different compared to the SI and TI measurement presented in ITU-T Rec. P.910 [3]. SI of a frame is measured by the standard division of each Sobel filtered frame in both vertical and horizontal direction as shown in Eq. 2.1.

$$SI_{Frame_i} = std_{space}[Sobel(Frame_i)]. \tag{2.1}$$

The TI for a frame is measured by the standard deviation over space of the difference ($D(Frame_i, Frame_{i-1})$) between corresponding pixel values in two adjacent frames (luminance component) as presented in Eq. 2.2

$$TI_{Frame_i} = std_{space}[D(Frame_i, Frame_{i-1})]. \tag{2.2}$$

Figure 2.9 illustrates the boxplot of SI and TI values for different gaming and non-gaming video sequences. It can be observed that, in general, SI for gaming videos has less variance compared to non-gaming videos. This might be due to the game's design, in which a game is created with a pool of predestined objects for which by the change of the scene over time, a large spatial similarity in terms

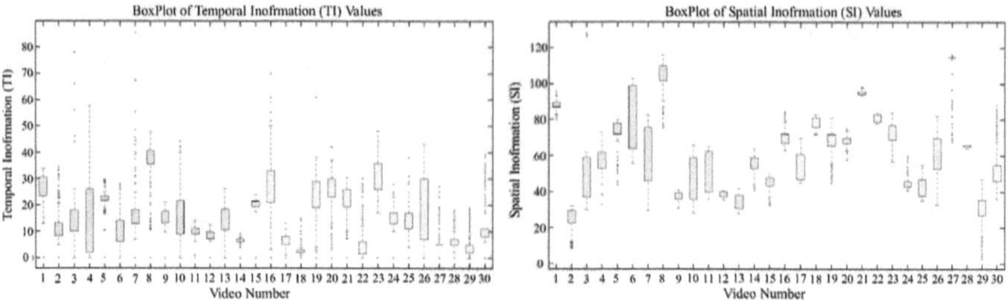

Fig. 2.9 Box Plot for SI and TI values of the 30 reference videos (Videos 1–15, Green: non-gaming videos; Videos 16–30, Cyan: gaming videos) based on [31]

of gaming objects can still be observed. In addition, each game has a certain level of abstraction, which is defined in the design process. These two particular characteristics of video games lead to less variation of SI values for video games compared to non-gaming content. In contrast to SI, considering TI values, gaming videos show slightly higher variance when compared to non-gaming videos, but no significant difference is observed.

As the next step, a total of 12 video sequences (six from each gaming and non-gaming groups) encoded using H.265/High Efficiency Video Coding (HEVC) at the native resolution of 1080p at five different CRF values (CRF = 26, 30, 34, 38 and 42, with lower CRF values implying better quality videos). Out of the six videos from each category, two videos each correspond approximately to low, medium, and high content complexity according to the SI and TI values. The video sequences are then used in a subjective experiment using Single Stimulus (SS) ACR on a 5-point scale, five corresponding to the best quality, according to ITU-T Rec. P.910 [3]. A total of 15 subjects participated in the test who rated the video quality of presented stimuli in the test. The ratings of all participants for each stimulus (each single encoded video sequence) is averaged to measure MOS values. The details about the subjective test and participants' demographic information are presented in [31], which for the sake of brevity are excluded from this section. In order to compare the performance of objective quality metrics on gaming and non-gaming content, four FR signal-based models are used for the encoded videos, Peak Signal to Noise Ratio (PSNR), Structural Similarity (SSIM) [72], Visual Information Fidelity - Pixel Domain (VIFP) [73] and VMAF [74].

To quantify the performance of the four objective metrics in relation to the subjective scores, the Pearson Linear Correlation Coefficient (PLCC) and Spearman's Rank Correlation Coefficient (SRCC) among the four objective metrics and the MOS values are measured, separately for the gaming and non-gaming videos. Table 2.4 presents the PLCC and SRCC values for all the four objective metrics. Based on the results presented in Table 2.4, the following observations can be drawn:

- As a general trend for both non-gaming and gaming videos, VMAF outperforms traditional objective metrics.

Table 2.4 PLCC and SRCC values of four objective metrics with respect to MOS scores

	PLCC		SRCC	
	Normal videos	Gaming videos	Normal videos	Gaming videos
PSNR	0.7499	0.6672	0.7496	0.7023
SSIM	0.7453	0.5781	0.8383	0.5991
VIFP	0.8085	0.7124	0.828	0.689
VMAF	0.9386	0.8954	0.942	0.8868

- All four objective metrics perform higher on non-gaming content compared to gaming content.
- Among the four evaluated metrics, SSIM performs the worst on gaming videos compared to non-gaming videos.
- While SSIM performs slightly better than PSNR for non-gaming content, this does not hold true for gaming content.

The results of this study showed that while the performance of FR quality models is found to be lower on gaming videos compared to non-gaming content, the difference is not too high to conclude that SoA FR metrics do not perform well on gaming content. In addition, it does not reveal the performance of other classes of quality metrics (e.g, NR metrics) on gaming content. Moreover, the videos are only encoded using a single resolution and CRF mode that does not represent the current encoding trends of gaming video streaming services such as cloud gaming and passive videos streaming, e.g., twitch.tv.

In the next section, another study is discussed in which the performance of 8 well-known FR, RR, and NR metrics on a gaming video quality dataset is evaluated under common encoding settings of passive gaming video streaming services.

2.7.3 Performance of Standard Image/Video Quality Models

The second study was designed and presented by Barman et al. [60, 61] in collaboration with the author of the book. The study aimed at the investigation of SoA image/video quality metrics' performance on a larger scale subjective test conducted solely with gaming video content. In the second study, a video quality dataset of gaming content was developed, called GamingVideoSet, that is described in detail in Sect. 3.4. The main differences between the two studies can be summarized as follows:

- A longer duration of 30-s stimulus (as suggested by ITU-T Rec. P.809) is selected instead of 10-s duration.
- As a choice of rate controller mode, CBR is selected following recommendations by various OTT service providers (e.g., twitch.tv).

Table 2.5 Comparison of the performance of the VQA metric scores with MOS ratings in terms of PLCC and SRCC values. All Data refers to the combined data of all three resolution-bitrate pairs. The best performing metric is shown in bold

		480p		720p		1080p		All Data	
Metrics		*PLCC*	*SRCC*	*PLCC*	*SRCC*	PLCC	SRCC	PLCC	SRCC
FR metrics	PSNR	0.67	0.64	0.80	0.78	0.86	0.87	0.74	0.74
	SSIM	0.57	0.43	0.81	0.78	0.86	0.90	0.80	0.80
	VMAF	**0.81**	**0.74**	**0.95**	**0.94**	**0.97**	**0.96**	**0.87**	**0.87**
RR metrics	ST-RREDOpt	−0.61	−0.51	−0.82	−0.85	−0.79	−0.92	−0.71	−0.74
	SpEEDQA	−0.63	−0.52	−0.83	−0.87	−0.77	−0.93	−0.71	−0.75
NR metrics	BRISQUE	−0.57	−0.48	−0.83	−0.89	−0.88	−0.91	−0.49	−0.51
	BIQI	−0.53	−0.51	−0.73	−0.72	−0.81	−0.80	−0.43	−0.46
	NIQE	−0.73	−0.74	−0.85	−0.81	−0.89	−0.90	−0.77	−0.76

- The videos are encoded under three widely-used resolutions, 480p, 720p, and 1080p, and rescaled to 1080p resolution for the subjective test.
- In addition to FR metrics, five well-known NR and RR models are also investigated in the study.

The performance of each image/video quality metric with respect to the subjective ratings is evaluated in terms of PLCC and SRCC in Table 2.5. Based on the results presented in Table 2.5, the following observations can be drawn:

- FR metrics generally follow a similar trend compared to the previous study, with one contradiction that SSIM performs significantly higher compared to the previous study.
- The performance of all eight metrics goes down at a low resolution of 480p.
- As expected, the FR metrics generally perform higher than RR metrics and RR metrics achieved higher performance compared to NR metrics.
- While the NR metrics performed reasonably on 1080p resolution, the performance of these metrics at a lower resolution is significantly decreased.

One of the main findings is the very low performance of existing NR quality metrics on gaming content. A possible reason behind such behavior could be due to the nature of the selected NR metrics. The NR metrics that were evaluated in the study were trained based on natural scene statistics. Thus, they might not be able to capture the combined effect of quality loss due to compression and quality loss due to rescaling. More recent studies confirmed the low performance of NR metrics on other gaming video quality datasets [50, 63, 75].

Based on the findings of the two studies that are described in this section, it can be concluded that while FR metrics perform reasonably on gaming content, there is a need for improvement of NR quality models to reach a satisfying performance on gaming videos.

2.8 Summary

In this chapter, an overview of the necessary steps in the quality assessment of the cloud gaming service is given. This includes primary knowledge about the gaming QoE and relevant quality features, identification of gaming-related influencing factors, information about the conduction of a subjective experiment for cloud gaming services, and an overview of related works with respect to the development of gaming quality models.

First, the concept of quality is described in the first section. Next, the QoE in the context of gaming is discussed based on the gaming QoE taxonomy. The taxonomy provides information about the quality features, influencing factors, and the relation between QoS and QoE aspects. While several features might influence the gaming QoE, for the development of a gaming QoE prediction model, it must be investigated based on the scope of the model and limitation of resources for subjective tests, timely and costly, which quality features could remain in the final prediction model.

In the third section, a detailed list of gaming QoE influencing factors that are grouped into human, system, and context factors is presented. The QoE influencing factors must be taken into account for the development of a prediction model. However, it is not possible to measure the effect of all mentioned influencing factors in this book due to a limited number of subjective tests that could be conducted. Thus, only the relevant factors that might be of interest to cloud gaming and network providers must be taken into consideration, and the remaining features must stay constant for all subjective tests.

Subjective testing in the context of cloud gaming services is described in the fourth section of the chapter with the necessary information that must be considered for planning and conducting the tests according to ITU-T Rec. P.809. Two paradigms of interactive and passive tests are introduced in this section, together with a brief description of a few questionnaires that could be used for assessing some of the gaming quality features. Based on the described procedure of subjective tests, it is decided to develop the gaming QoE model based on datasets created in both passive and interactive paradigms. The passive paradigm allows collecting a large number of ratings to assess video quality and test more encoding parameters and a higher range of parameters, which is not possible in the interactive test paradigm due to a longer test duration. However, parameters that could influence the interaction of players to the system (e.g., delay) are evaluated in the interactive test.

Finally, the chapter ended with an overview of related works that focus on development of quality prediction models for gaming streaming applications. Based on this section, it can be concluded that, up to now, the research works that have been done with respect to the development of gaming quality models in the literature are limited to small scale subjective tests and a limited number of influencing factors with a lack of any comprehensive model that could be used for quality prediction of cloud gaming services. In addition, even the SoA video quality metrics with many years of research, are still not capable of reaching a reasonable performance for gaming video content, especially for NR metrics.

In this book, multiple subjective tests in both interactive and passive paradigms are conducted on a large scale, according to ITU-T Rec. P.809 based on a large number of influencing factors. A gaming QoE model is developed that can be used for different purposes such as network planning, resource allocation, and monitoring of quality for cloud gaming service. Details about the model structure are presented in the next chapter.

Chapter 3
Process for Model Development

3.1 Structure of the Framework

Gaming QoE is a multidimensional construct in which many quality features have an impact on the overall QoE. However, in order to develop a quality model that can be used by network providers or service providers, it is not realistic and practical to measure all the quality features such as immersion, presence, and flow, since the estimation of these quality features requires having access to user experience, user preference, and information about the game design which are not always available to network or service providers as discussed in Sect. 2.3. Therefore, to develop the gaming QoE model, two important quality features, video quality and input quality are considered to predict the cloud gaming QoE. It has to be noted that the decision to take into account the video and input quality features is not only based on the level of access to the information but also based on the structural equation modeling, i.e. a kind of regression analysis, that has been conducted by Schmidt [48] on a large interactive dataset using the ratings of multiple gaming quality features as the independent variable and overall gaming QoE ratings as the dependent variable. The result presented by Schmidt showed that 62% of the variability in the gaming QoE ratings could be explained only by the subjective ratings of GIPS items, i.e., represent input quality, video quality, and video discontinuity. While this result is based on the individual ratings of players, if the MOS values are used, the regression achieves an R-squared of 0.92. If other quality features related to player experience are taken into account, only 9% higher variability in the gaming QoE ratings (71%) can be explained by the model. The details of this analysis can be found in [48].

The framework to predict cloud gaming quality is illustrated in Fig. 3.1. The model framework consists of two modules of input quality and video quality. Video compression and network transmission error impact the video quality, while end-to-end delay, as well as transmission error, affect the input quality.

S. Zadtootaghaj, *Quality of Experience Modeling for Cloud Gaming Services*,
T-Labs Series in Telecommunication Services,
https://doi.org/10.1007/978-3-030-98249-2_3

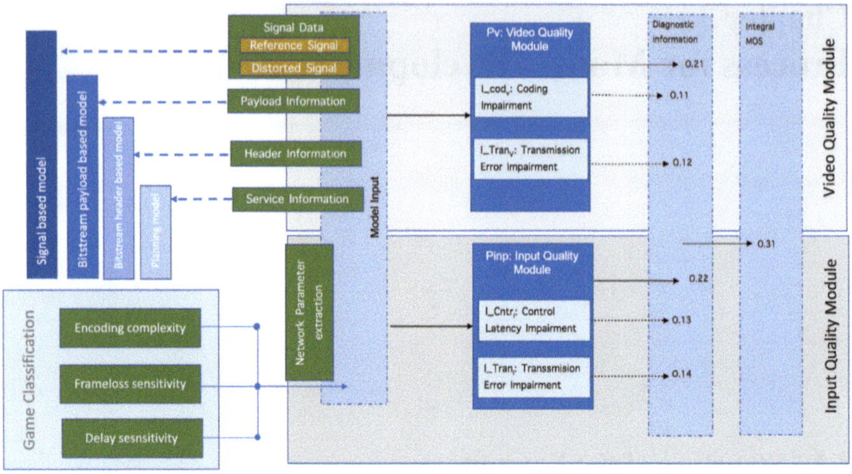

Fig. 3.1 Proposed structure of the model to predict the gaming QoE

The video quality can be predicted with different types of models depending on the level of access to the information, computation requirement, and purpose of the quality assessment as discussed in Sect. 2.7.1. Thus, different types of models are developed to predict the video quality that is degraded by compression, e.g., bitstream-based and signal-based models.

The game classification is an essential component to obtain a precise prediction for video games that is discussed in Sect. 3.6. The impact of transmission and encoding distortions on various quality features perceived by a player depends strongly on the sensitivity of a game towards these degradations. In order to reach a higher accuracy of the model, two modes of operation are defined depending on whether a network planner or service provider can make any assumption on the type of game that is targeted or not. If the network planner or cloud gaming provider has knowledge or assumption about the type of the targeted game with respect to its encoding complexity and sensitivity towards delay and frame losses, the gaming QoE can be predicted more accurately by considering the game classes. If the user of the developed models has no knowledge or assumption on the game type, the highest content class will be assumed as a default mode. Therefore, in the default mode, the model will result in a pessimistic quality prediction for games that are not highly complex and sensitive. Considerations on how to quantify game characteristics to derive the classes considered for the model are given in Sect. 3.6.

At the end of the building block, two layers of diagnostic information and integral MOS are placed. Diagnostic information gives insight into the degradation type and can be measured separately without testing the whole model. For example, the impairment due to video encoding is an important factor that might be desired to be predicted for other services such as passive video gaming streaming, e.g., twitch.tv. Integral MOS (also named as core model in later chapters) is responsible for the integration of all measured impairment factors to predict the final output quality (O.31).

Finally, the model returns multiple outputs, in a 100-point scale that is referred to as "R-scale" in ITU-T Rec. G.107 [66], from the diagnostic layer and integral layer as listed below:

- $O.31$ is the estimated gaming QoE expressed on the R-scale, where 0 is the worst quality and 100 the best quality.
- $O.21$ is the estimated video quality, impaired due to video compression artifacts as well as transmission error on the R-scale.
- $O.22$ is the estimated input quality, impaired due to control latency impairment and transmission errors on the R-scale.
- $O.11$ is the estimated video coding (compression) impairment affecting the visual perception on the R-scale.
- $O.12$ is the estimated transmission error impairment affecting the visual perception on the R-scale.
- $O.13$ is the estimated control latency impairment affecting the input quality on the R-scale.
- $O.14$ is the estimated transmission error impairment affecting the input quality on the R-scale.

3.2 Model Development

For modeling, the impairment-factor based approach is followed according to the Allnatt [76] findings. Generally, the proposed models follow the core structure of Eq. 3.1.

$$Q_X = Qo_X - \sum_{k=1}^{N} IF_k^X \tag{3.1}$$

where Q_X is audio, video, audiovisual, or input quality. In this book, the audio is not considered in the model development; thus, Q_X only represents video and input quality. Qo_X is the base video, or input quality, and takes the maximum value of the modeling scale. N represents the total number of Impairment Factors (IF), and IF_k^X denotes an IF for modality X with indices k.

Three impairment factors are used for the video and input quality models. The first one relates to the quality impact due to compression artifacts which is investigated in the video quality component of the framework. The second impairment factor is the effect of network transmission errors on the input quality and video quality. Finally, the quality of the control stream could be impaired due to network latency. Control stream is referred to the uplink and downlink data transmission that the desired action of players through capturing the players' input (keystroke) is transmitted over a broadband network to the cloud gaming server until the result of the requested action is displayed on the player's screen. The core structure can be extended according to the different types of impairment following Eq. 3.2.

$$Q_X = Qo_X - I_{cod_X} - I_{trans_X} - I_{cntr_X} \tag{3.2}$$

The impairment factor I_{cod_X} impacts the video quality solely, while I_{trans_X} can affect both input and video quality features. Transmission errors due to packet loss are typically concealed by dropping frames (named as freezing effect, which is discussed in Chap. 5) in cloud gaming services. Low displayed framerate (either due to encoding or packet loss) does not only impact the visual perception (video quality) but also the interaction of the player with the game (input quality), e.g., the controllability and responsiveness aspects (cf. Sect. 2.5.5). The control stream can be impaired due to the latency of multiple processes between client and server such as data propagation, packet transmission delay, and encoding processing. This latency affects strongly the interaction of the players with the game which is reflected in the input quality.

The core model can be further extended based on the input quality and video quality modules according to Eq. 3.3.

$$Q_{Gaming} = Qo_{Gaming} - a \cdot I_{cod_V} - b \cdot I_{trans_V} - c \cdot I_{trans_I} - d \cdot I_{cntr_I} \tag{3.3}$$

In addition, if the multidimensional approach for impairment factor I_{cod_X} is followed, the core model can be extended upon the sub-dimensions of video quality, video fragmentation (VF), video unclearness (VU), and video discontinuity (VD) as shown in the Eq. 3.4.

$$Q_{Gaming} = Qo_{Gaming} - a \cdot I_{trans_V} - b \cdot I_{trans_I} - c \cdot I_{cntr_I} - d \cdot I_{VU} - e \cdot I_{VF} - f \cdot I_{VD} \tag{3.4}$$

The impairment-factor-based video quality and input quality models are described in Chaps. 4 and 5 accordingly.

3.3 Model Training

The gaming QoE prediction models are developed based on the datasets described in Sect. 3.4. In total, four gaming quality datasets, including two video quality datasets, an interactive dataset, and an image quality dataset, are created during the research of the present book in collaboration with other researchers.

For all video quality datasets, a typical 5-point scale ACR with hidden reference (ACR-HR) is targeted, but for one of the video quality datasets, a 7-point EC scale is used (according to ITU-T Rec. P.809 [14]), which latter the MOS values are transformed back to the 5-point ACR scale to be in-line with other datasets. Based on Allnatt's approach [76], the models are trained either based on a 100-point scale, or 5-point MOS scale. This 100-point scale is referred to as "R-scale" in ITU-T Rec. G.107 [66]. The process of transforming the EC scale to the ACR scale and finally to the R-scale together with the data cleansing process is discussed in detail in Sect. 3.5.

In the training process, the passive video quality datasets are used to develop models that can predict the impairment of video coding, I_{cod_V}, while other impairment factors are trained based on the collected data in the interactive dataset. However, a few encoding conditions are also tested in the interactive test in order to merge the result of passive and interactive datasets. This separation of development is done based on a number of reasons. First, this separation allows collecting more data to predict the impairment of video coding due to reduction of questionnaire items in the passive viewing-and-listening tests as well as shorter duration of each stimulus, 30 s, compared to 90 s in the interactive test. Collecting video quality ratings through a large passive viewing-and-listening test allows developing more accurate video quality models, which were not possible through the interactive studies with the same available resources. Second, the passive viewing-and-listening tests ensure that all participants are exposed to the same audiovisual content, whereas in the interactive test, players have different experiences depending on their gameplay.

3.4 Database Overview

During the research of the present book, multiple datasets are developed in collaboration with other researchers, which allow the development of QoE prediction models for gaming streaming services. In this section, an overview of five quality datasets is given. Among them, three datasets are developed based on passive viewing-and-listening quality assessment tests, which the author majorly contributed to two of them, the Gaming Video Dataset (GVSET) and the Cloud Gaming Video Quality DataSet (CGVDS). GVSET and Kingston University Gaming Video Dataset (KUGVD) are built by encoding videos using a software implementation of H.264/MPEG-AVC. However, for delay-sensitive cloud gaming services, currently, most of the providers (e.g., Geforce Now and Parsec) use the hardware-accelerated implementation of the H.264/MPEG-AVC standard. Due to the lack of available datasets using such a fast encoding setting, CGVDS was created in which the videos are encoded using the hardware-accelerated implementation of H.264/MPEG-AVC (NVENC), with also a wider spread of video games and more encoding parameters. Table 3.1 gives a comparison of the three video quality datasets.

In addition to the gaming video quality datasets, an image quality dataset was created using gaming images. Finally, a large interactive quality assessment experiment was conducted to develop QoE models for cloud gaming services. This dataset was initially developed under ITU-T work item G.OMG which led to ITU-T Rec. G.1072.

Table 3.1 Comparison of existing gaming video quality datasets

	GVSET [77]	KUGVD [63]	CGVDS [50]
Influencing factors	Resolution, bitrate	Resolution, bitrate	Resolution, bitrate, framerate
Framerate	30 fps	30 fps	60, 30, 20 fps
Preset	Veryfast	Veryfast	llhq
Number of stimuli	90	90	380
Encoder	FFmpeg, x264	FFmpeg, x264	NVENC (H.264)
Encoding mode	CBR	CBR	CBR
Resolution	480p, 720p, 1080p	480p, 720p, 1080p	480p, 720p, 1080p
Questionnaire items	VQ, Acp[a]	VQ, Acp	VQ, VF, VU, VD, Acp

[a] Acp stands for acceptance of the service

Table 3.2 Resolution-bitrate pairs used to obtain distorted (compressed) video sequences. The bitrates in bold text refer to the bitrates used in the subjective quality assessment

Resolution	Bitrate (kbps)
1080p	**600**, **750**, 1000, **1200**, 1500, **2000**, 3000, **4000**
720p	**500**, **600**, 750, 900, **1200**, 1600, **2000**, 2500, **4000**
480p	**300**, 400, **600**, 900, **1200**, **2000**, **4000**

3.4.1 GVSET

The GamingVideoSET, [77], which is referred to as GVSET in the remainder of the book, consists of 24 source video sequences, each 30 s duration. The videos are recorded losslessly at a framerate of 30 fps from 12 different games where two sequences per game are recorded. The reference videos are encoded using H.264/MPEG-AVC under 24 multiple resolution-bitrate pairs resulting in a total of 576 distorted sequences. In addition, a subjective test considering 90 impaired sequences of this dataset has been conducted and released with the dataset. The subjective ratings are collected in a 5-point ACR scale for video quality as well as acceptance of the presented visual stimulus. For all encoded videos, multiple objective metrics are reported, e.g., PSNR, VMAF. Table 3.1 presents a summary of encoding settings chosen for GVSET and Table 3.2 describes the resolution-bitrate pairs considered in the dataset. Figure 3.2 presents the MOS of the video quality ratings for the six selected gaming videos of the subjective test at 480p and 1080p resolutions.

3.4.2 KUGVD

The Kingston University Gaming Video Dataset (KUGVD) consists of six high-quality raw gaming videos of each 30 s duration similar to GVSET, at 1080p

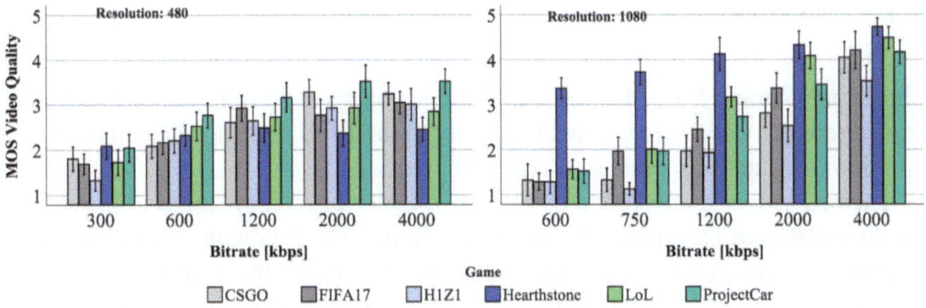

Fig. 3.2 Barplots of the mean video quality scores and 95% confidence interval for the six selected games for the used bitrates at two different resolutions of 480p (left) and 1080p (right) of GVSET dataset

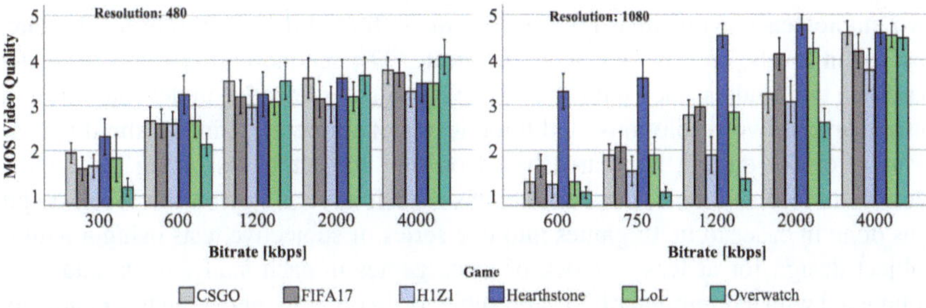

Fig. 3.3 Barplots of the video quality scores and 95% confidence interval for the six selected games for the used bitrates at two different resolutions of 480p (left) and 1080p (right) of KUGVD dataset

resolution, and a framerate of 30 fps [63]. Subjective quality ratings are collected for 90 stimuli from encoded videos using the H.264/MPEG-AVC codec standard in 15 different resolution-bitrate pairs similar to GVSET. Additionally, for all encoded videos, VMAF ratings, as well as some other well-known objective metrics, are available. The resolution-bitrate pairs and encoding settings follow the GVSET; thus, the reader can refer to Tables 3.2 and 3.1 for detailed information about the resolution-bitrate pairs selected for the subjective test and the encoding settings of KUGVD. In addition, two bar plots of participants' ratings for resolutions of 480p and 1080p are illustrated in Fig. 3.3. It has to be noted that while five out of six selected games of KUGVD are the same as GVSET, the recorded video sequences are different.

3.4.3 CGVDS

The Cloud Gaming Video DataSet (CGVDS) consists of 15 raw recorded gameplays of different games, each recorded losslessly in the RGB format at 1080p resolution, 30 s duration, and framerate of 60 fps [50], using Fraps.[1]

The raw videos are encoded by FFmpeg[2] under different resolution-bitrate pairs for three levels of framerate, 20, 30, 60 fps. In order to follow the encoding settings of cloud gaming services, all videos are encoded using the hardware-accelerated implementation of H.264/MPEG-AVC (NVENC) using llhq preset. The llhq preset does not use B-frames and it uses an infinite GoP length. Due to a large number of stimuli for subjective tests and to avoid fatigue during the test, the experiment was split into three parts, and five different studies, using 72 stimuli per study (except one study that 92 stimuli were used). The first part (Part-1) consists of two studies each with three raw video games, four different bitrate levels, three resolution levels, and two levels of framerate. The second part (Part-2) is created based on two studies, each with three raw video games encoded under four different bitrate levels, two resolutions, and three levels of framerate. Finally, the third part (Part-3) of the dataset is created based on one subjective test using three video games, three resolutions, three framerates, and four levels of bitrate. Such a split was done in order to fit 15 games into five series of subjective tests using a within-subject design for at least a block of three games in each study. Each study was conducted with a minimum of 20 valid subjects. In order to merge the five conducted studies, three video sequences as anchor conditions were added to the subjective test, one sequence with low bitrate triggering the fragmentation dimension, one with low resolution triggering the unclearness dimension, and finally a reference video sequence. Moreover, prior to the test, three training video sequences are shown to the participants for training purposes and to ensure that the participants understand the used measurement scales (cf. Sect. 2.5.2).

A large subjective experiment is conducted to assess the video quality of several stimuli, following the ITU-T Rec. P.809 [14]. In total, 380 stimuli are used in the whole experiment in which over 150 participants rated the video quality and four other post-condition items. As the choice of the post-condition questionnaire, five items are selected including overall video quality, three perceptual sub-dimensions of video quality proposed in ITU-T Rec. P.918 (cf. Sect. 2.5.5.2), as well as an acceptance of video quality, have been assessed. Following the ITU-T Rec. P.809, instead of the 5-point ACR scale that is typically used for video quality assessment, the 7-points EC scale is used for the experiment. At the end of the experiment, a few post-test questions about the importance of different quality dimensions are asked in order to get an insight into the judgment process of participants. To get more information about the dataset and experiment methodology, please refer to [50].

[1] https://www.fraps.com.

[2] https://ffmpeg.org.

3.4.4.2 Test Methodology

In order to get insights into the influence of encoding parameters on the perceived video quality, in addition to overall image quality, and with inspiration from video quality dimensions (cf. Sect. 2.5.5.2), two more questions are added in order to assess the fragmentation and unclearness. While this method was proposed for a video quality assessment, it is found to be a suitable method to identify the types of degradation for image content as well.

Each dimension is explained to the participants in an introduction session in a written form using describing adjectives and in the form of example images. As described in Sect. 2.5.5.2, the rating scales for DBSQE-V are designed based on a 7-point continuous scale, using the antonym pairs to describe the range of the scales. However, the 5-point ACR scale for overall image quality, fragmentation, and unclearness is used in order to be consistent with typical image quality tests as well as with the ratings of GVSET.

Since this method is not validated for image content, the scatter plots of overall quality, fragmentation and unclearness are drawn in Fig. 3.6 to get more insight into the suitability of the method. The scatter plot 3.6c reveals a high difference between fragmentation and unclearness ratings. This indicates that participants can clearly differentiate between fragmentation and unclearness dimensions. To ensure that the difference between fragmentation and unclearness is meaningful and not due to the vagueness of the added dimensions, a Multiple Linear Regression (MLR) model is fit based on the predictors image fragmentation and image unclearness to predict the overall image quality. The MLR model predicts the overall image quality with a PLCC of 0.98 and RMSE of 0.154, based on Eq. 3.5. These dimensions can be used as a diagnostic approach to spot the reasons behind the low image quality. Figure 3.6d shows how well the MLP model can predict the image quality based on the fragmentation and unclearness ratings.

$$IQ_{Estimated} = -1.073 + 0.657 \cdot VF + 0.573 \cdot VU \tag{3.5}$$

Finally, the scatter plot reveals a proper distribution of ratings from both sides of the scale except for the small area between 4 to 4.5 MOS. If the subjective rating distribution would be skewed to one side of the scale then there might be a bias on the training process.

3.4.5 Interactive Dataset

In order to develop a model that can predict cloud gaming quality, a dataset developed following the interactive paradigm is required. Thus, an interactive dataset is developed using an actual cloud gaming setup and following the ITU-T Rec. P.809 interactive test paradigm. The test setup requires four essential elements as described below as well as in Fig. 3.7:

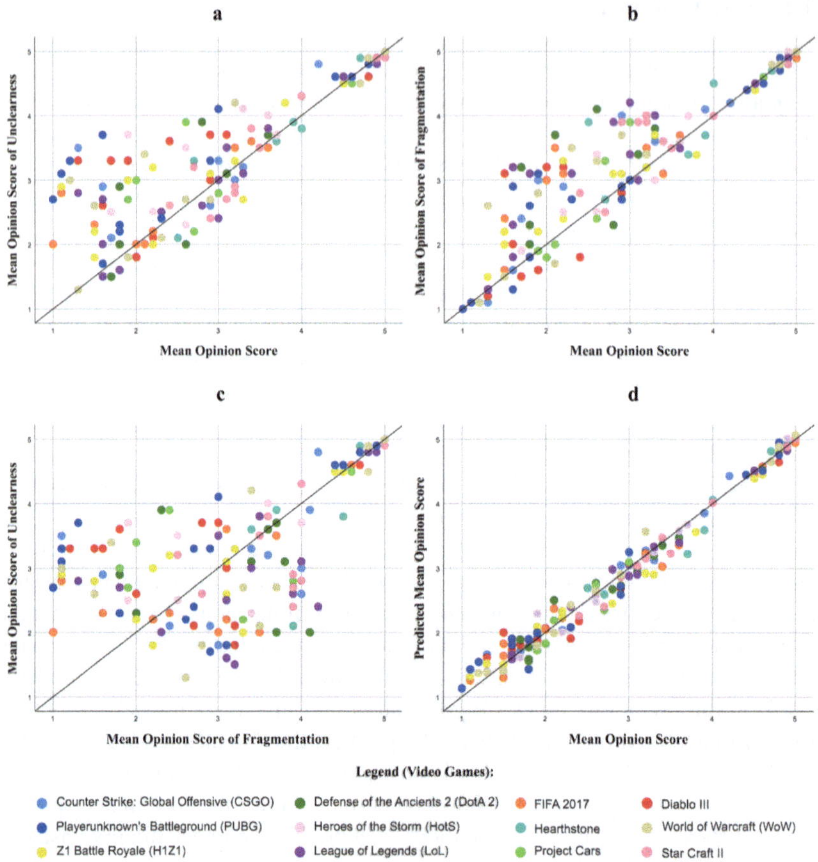

Fig. 3.6 The scatter plot of ratings distribution for different quality dimensions as well as image quality prediction. (**a**) MOS vs Unclearness. (**b**) MOS vs Fragmentation. (**c**) Fragmentation vs Unclearness. (**d**) MOS vs Predicted MOS

1. A server PC that is powerful enough to run the selected games without any issues such as variable framerate due to low processing power.
2. A client PC capable of playing the video stream smoothly.
3. A local network for controlled streaming conditions.
4. Peripheral components such as monitors, headset, and input devices.

To change the encoding or network conditions, a script was implemented that controlled all settings on the client PC. As a test environment, lab rooms offering high control over lighting during a test were selected, adhering to ITU-T Rec. P.910 [3]. In addition to the components listed above, a laptop was used, which runs a digital questionnaire that participants filled out after each condition. The test instructions, a pre-test, post-game, and post-test questionnaires are provided on this laptop. The structure of a subjective study is shown in Fig. 3.8. In each study, one

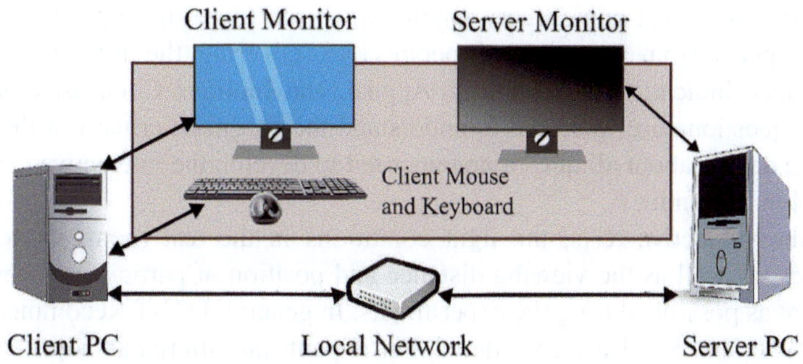

Fig. 3.7 Setup for interactive subjective tests

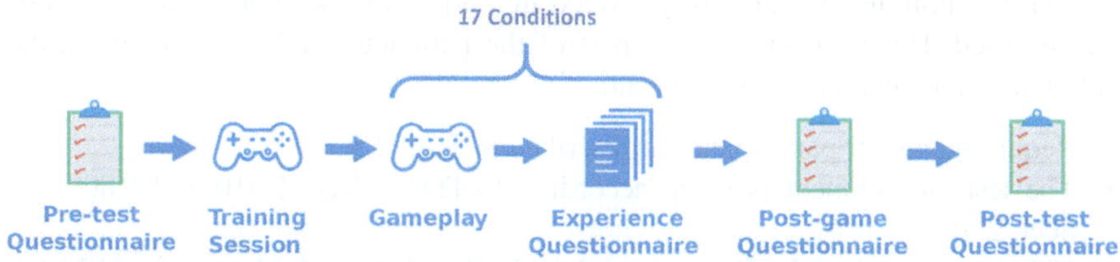

Fig. 3.8 Structure of subjective test for interactive tests

game is tested under 17 different conditions in which the bitrate, delay, packet loss, and encoding framerate are changed.

Three types of questionnaires are used in the interactive test. Prior to the experiment, the demographic information is collected using a pre-test questionnaire which covers the information about age, gender, gaming experience, gaming preference, knowledge about the games used in the study, and commonly used input device and monitor. The demographic information collected in the experiment is given in Appendix A.

In addition, a 31-item post-condition questionnaire (used after each stimulus) is used to assess the gaming QoE of the presented stimuli. Participants are asked to indicate their experience of their play for multiple items on the 7-point EC scale. The post-condition questionnaire covers the following aspects:

- Overall gaming QoE
- Input quality based on the GIPS questionnaire (cf. Sect. 2.5.5)
- Output quality: Video quality, sub-dimensions of video quality (see Sect. 2.5.5.2), audio quality.
- Player experience (PX) based on the iGEQ questionnaire [12] covering seven dimensions of Competence, Immersion, Flow, Tension, Challenge, Positive and Negative Affects.
- Player Performance and Service Acceptance.

In addition, a post-game questionnaire was used to get insight into the game-related aspects from participants' perspective, covering the following aspects: Performance Indication, Learnability, Appeal, and Intuitive Controls. Finally, the post-test questionnaire was used to understand the judgment criteria of the participant. The details about all questionnaires used to develop the interactive dataset are provided in Appendix A.

For the subjective tests, the light conditions in the test room, its acoustical properties, as well as the viewing distance and position of participants were set as consistent as possible during the experiments. In general, ITU-T Recommendations P.809, P.910, and P.911 are considered. Participants are offered an adjustable chair and table to sit in a proper position. The viewing distance D is set to equal to three times the picture height H (the video window size, not the physical display size), D = 3*H. For both test paradigms described in the previous section, different setups can be used. However, unless it is part of the parameter under investigation, the following elements remained constant:

- Requirements for participants as described in Sect. 2.5.
- The test environment is setup according to ITU-T Rec. P.910 and P.809 (cf. Sect. 2.5).
- Video presentation: LCD monitor, 24" display for standard gaming, HD1080 resolution, no G-Sync, no Free-Sync, (others see [b-ITU-T P.911]).
- Input device: only a mouse and keyboard are used.
- Play audio using a diotic presentation (both ears receive the same mono signal) or binaural presentation (each ear receives one channel of a stereo signal).

In total, nine studies are conducted that for each subjective test, 30 participants were invited who should fulfill some criteria, concerning gaming experience, normal color vision and visual acuity, a fair language skill, and having no relevant neurological disease (e.g., epilepsy) or sensory-motor dysfunctions.

An extensive subjective test following the interactive paradigm guideline described in Sect. 2.5.3 is conducted. The interactive dataset contains the subjective ratings of a total of 180 participants. The data was collected from a total of 9 studies, each testing a single game. Each study lasts a maximum of two hours using a total of 17 conditions. Four types of degradation are introduced, including round-trip delay, encoding framerate, encoding bitrate, and packet loss that results in frame loss (cf. Chap. 5). Table 3.4 illustrates the range of values for factors and parameters in the interactive dataset. For the shooting game, Counter-Strike: Global Offensive (CSGO), an additional experiment was conducted to investigate the relationship between spatial video quality and input quality.

Figures 3.9 and 3.10 present the barplots for a few conditions and games tested in the interactive dataset for variable framerate, bitrate, delay and frameloss rate. As it can be observed, the games are not equally affected by introduced degradation. While some games are strongly impacted by degradation, such as network delay, others seem to be less sensitive to delay. Such a difference in the effect of degradation on participants' ratings offers challenges in the development of gaming QoE models that rely solely on parameters from network and compression. Therefore, in

Table 3.4 Factors and parameters used in interactive tests

Application information	Value range, unit
Sequence duration	90 s
Screen size	24
Input devices	Mouse and keyboard
Packetization	RTSP (over RTP/UDP/IP)
Video codec	h.264 using NVENC
Resolution	1080p (*480p also tested only for CSGO*)
Coded video bitrate (mbps)	2, 4 ,6, 50
Frame rate (fps)	10, 20, 30, 60
Group of pictures (Note 1)	Infinite
Pre-set	llhq (low latency, high quality)
Encoding mode	CBR
Video compression	Standard H.264, Main 4.0
Audio codec	AC3
Coded audio bitrate (kbps)	192 (stereo)
Audio sample rate (Hz)	48,000
Packet loss degradation	Uniform loss (0–5%)
Freezing ratio due to packet loss	0, 10, 20, 30, 40, 50%
Delay range	0, 25, 50, 100, 200, 400 ms

Note 1: The llhq preset does not use B frames and it uses an infinite GoP length by default. In case of a corrupted frame, no spatial artifacts will be visible but instead the FEC will lead to a replacement of the corrupted frame resulting in jerkiness of the video (freezing artifacts)

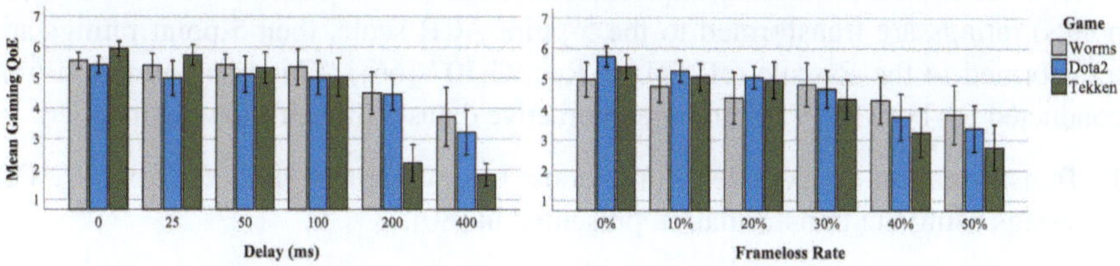

Fig. 3.9 Bar plots of mean opinion scores of gaming QoE with 95% confidence interval of two network parameters, delay and frameloss rate, for interactive tests for three different games

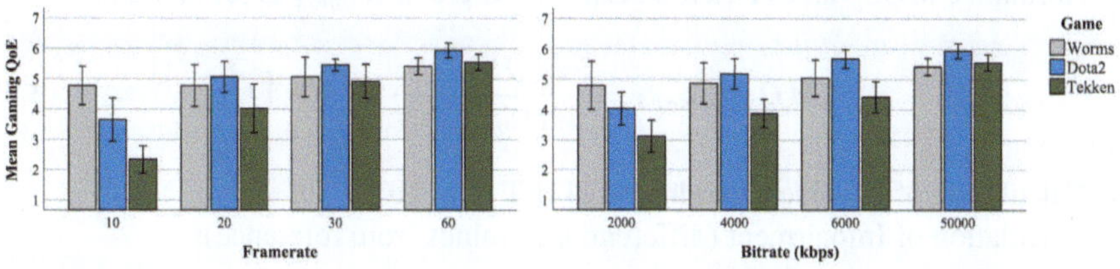

Fig. 3.10 Bar plots of mean opinion scores of gaming QoE with 95% confidence interval of two encoding parameters, bitrate and framerate rate, for interactive tests for three different games

the following sections, three different game classifications are proposed to group the games based on the characteristics that cause them to be sensitive towards a certain type of degradation.

3.5 Data Post-Processing

Once the ratings are collected from each described experiment, the cleaning process is applied to fill the missing values and remove the outliers. The outlier detection and removal process was applied to the raw subjective scores according to the outlier labeling rule presented in [79] for each test condition of the dataset. In addition, values with a higher distance than 2.2 interquartile range from the median of ratings were marked as outliers. In addition to the typical outlier detection, in the interactive dataset, it was observed that 24 participants rated the quality of the reference stimuli lower than 4 on a 7-point EC scale. Thus, all ratings of these participants are removed from the dataset. It has to be noted that the low ratings of participants for the reference condition might be due to their preference in a certain genre or low performance in the played condition. However, for the modeling purpose, this might introduce additional errors to the model since the model does not take into account the quality features that reflect such a user preference or performance indication. After removing the outliers from the initial collected 2960 ratings, 2648 ratings remained for further analysis.

In order to derive suitable impairment factors for the model, the EC scale (7-points) ratings are transformed to the 5-point ACR scale, then 5-point ratings are transformed to the R-scale (cf. ITU-T Rec. G.107 [66]). This transformation was conducted for both CGVDS and the interactive dataset in the following manner:

1. Transformation of extended continuous 7-point ratings (EC) to 5-point ACR ratings using the transformation presented in [80].

$$\widehat{MOS_{ACR}} = -0.0262 \cdot \widehat{MOS_{EC}^3} + 0.2368 \cdot \widehat{MOS_{EC}^2} + 0.1907 \cdot \widehat{MOS_{EC}} \cdot +1 \tag{3.6}$$

2. Normalize MOS values based on Eq. 3.7, where MOS_{max} is set to 4.64.

$$MOS_{normal,i} = \frac{MOS_i - 1}{MOS_{max} - 1} \cdot 3.5 + 1 \tag{3.7}$$

3. Calculation of R-value according to ITU-T Rec. 107.
4. Calculation of Impairment (differential R-values from reference):

$$I_{factor}(condition) = R_{max,factor} - R_{factor}(condition) \tag{3.8}$$

The derived $I_{factor}(condition)$ for quality features used in the dataset such as video quality and input quality (based on GIPS items) are used for training the model. For the sake of simplicity, the $I_{factor}(condition)$ is referred to as delta-R in the remainder of the book.

3.6 Gaming Classification

In several studies, it has been shown that video quality [30] as well as gaming QoE [28] are strongly content dependent. The game has a strong impact on the perceived video quality and interaction of players with the game. This can be seen from the bar plots of Figs. 3.9 and 3.10 in which within the same condition, the subjective ratings of the overall quality vary strongly depending on the game.

The commonly used game genre classification has been shown to be inaccurate in predicting the game sensitivity towards different types of impairments. For example, Schmidt has shown in a study that the games within the same genre might have different sensitivity towards delay [28]. Besides, due to the overlapping of genres, this classification is not considered a reliable tool to classify the games upon their sensitivities towards network and compression impairments.

Due to the strong influence of the game design on perceiving different types of degradation, it is not possible to develop an accurate quality prediction model without considering the game sensitivities/complexities. Therefore, three types of gaming classifications are taken into consideration when assessing and predicting gaming QoE. First, a classification that estimates the video complexity based on certain game characteristics. While the signal-based and bitstream-based video quality models have enough information to estimate the video complexity, the planning models' performance relies strongly on such a video complexity classification due to a very low level of information that they have access to. Second, a delay sensitivity classification is proposed to classify the games based on their sensitivity towards delay degradation. Finally, the degradation due to a low encoding framerate and frame loss due to packet loss does not influence the interaction of players similar to delay degradation, and hence a frame loss sensitivity classification is required as well.

In the following, a short overview of the development process of these three classifications is given. This section is mainly written based on two publications of [30] as well as [81].

3.6.1 Gaming Video Complexity Classifications

In contrast to traditional video content, gaming content has special characteristics such as an extremely high motion for some games, special motion patterns, synthetic content, and repetitive content, which makes the state-of-the-art video and image quality metrics perform poor for this special computer-generated content [30, 31]. Video games are usually created based on a pool of limited pre-designed objects that appear in different scenes of a game. Therefore, there is a high similarity of different scenes of a particular game with respect to the spatial video complexity (e.g., level of texture). Besides, for many video games, due to the same design style across the games, a game shares similar visual features such as background scene, color diversity, and pattern of motion [30]. Moreover, for many games, the game world is small, and the game scenes do not change much over time. Thus, these characteristics of the games allow concluding that for most of the video games, a level of video complexity can be assigned to the game. It has to be noted that such a class assignment would only target the representative scene and action of a game that frequently happens in a certain game, and the video quality of the game strongly depends on these scenes. For example, in a shooting game, a representative scene could be a scenario that the main character is in a shooting task and not when the main character stands in a position and zooms his sniper on enemies.

In order to develop a gaming video complexity classification, two studies are conducted. The first one was conducted in early 2018 when there was no large gaming quality dataset available. Thus, instead of using subjective ratings, an objective video quality metric, VMAF, which has been shown to correlate well with subjective ratings, was used for the classification. Another study was conducted in late 2019, where the ratings from the interactive dataset were used as a ground truth [81].

In the first attempt [30], to derive a variety of game characteristics relevant to the video coding complexity of gaming content, a two-step focus group interview with three experts was conducted. Among the three experts, two of them are gaming experts and have a fair knowledge of video encoding. One of them is an expert in the video domain with a reasonable knowledge of gaming. The three experts were kept the same for all analyses. In the first step, various game video scenes (mostly from GVSET) at 1080p resolution encoded with a bitrate of 1 Mbps were presented to the experts. The bitrate of 1 Mbps was selected to have a comparative range of visible blockiness among the different gaming videos, which can then be used to define the game characteristics with a possible influence on the encoding complexity. The experts were instructed only to define characteristics that are visually quantifiable by someone with reasonable domain knowledge. As an outcome of the first session, the experts provided a list of nine game characteristics that could potentially be a reason for the difference in the amount of blockiness among the games. Finally, the characteristics are grouped according to their possible influence on the video scenes' temporal and spatial complexity. In the second step of the interview, the same experts are asked to describe the means to quantify the characteristics and assign values for each characteristic of the game videos. During the second interview, the experts were asked to quantify the different characteristics of the 30 videos of the database. Once the game characteristics are quantified, they should be mapped to different levels of video complexity.

Therefore, in the next step, video game complexity is estimated based on Rate-Distortion (RD) curves. In order to build the RD curves, the raw recorded gaming scenes at 1080p resolution and framerate of 30 fps are encoded using the H.264/AVC codec using CBR rate controller model, in the bitrate rang of 750 kbps

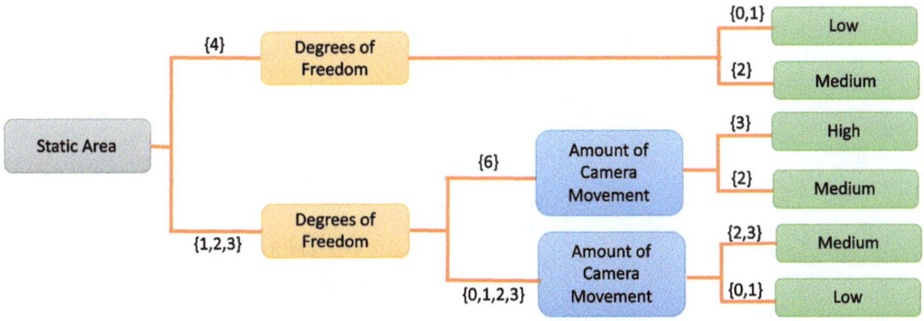

Fig. 3.11 Decision tree determining the encoding complexity of a game scene (cf. [30])

to 4000 kbps. The RD curves are plotted according to bitrate level and VMAF values. Next, a k-means clustering algorithm was performed to cluster the 30 video sequences into different groups based on their RD curves. The optimum number of clusters was found to be three based on a silhouette analysis (silhouette value = 0.84). After clustering the quality values into three groups, in order to map the game characteristics to the clusters, a decision tree was achieved based on the assigned game characteristic values as illustrated in Fig. 3.11. The best possible decision tree was achieved with an accuracy of 96% using three characteristics described below.

(a) **Static Areas (SA):** A static area is a part of the scene without any movement. This typically applies to maps, background, and Heads-up display (HUD) showing skills or scores. To quantify this characteristic, the size of the static areas in relation to the overall video size should be estimated as an average over time.
(b) **Degrees of Freedom (DoF):** Degrees of freedom is defined as the freedom of camera movement. There can be up to six degrees of freedom due to three possible translations (back and forward, left and right, or up and down) as well as three rotations (vertical axis and height) of the camera.
(c) **Amount of Camera Movement (ACM):** How often the camera is moving plays an important role in the temporal complexity of a video. Differences are to be expected when comparing turn-based, and real-time games [82]. Experts stated that they expect strong variations for this characteristic, even for the same game, depending on the game scene selected. Due to the importance of scene selection, scenes that do not cover the typical behavior in the game should be avoided for building a model or classification. The amount of camera movement in this work is defined as the percentage of the total duration of camera movement to the overall duration of the game scene. Depending on the percentage of camera movement, four categories were assigned: 0–10% (0), 10–49% (1), 50–99% (2) and 100% (3) camera movement.

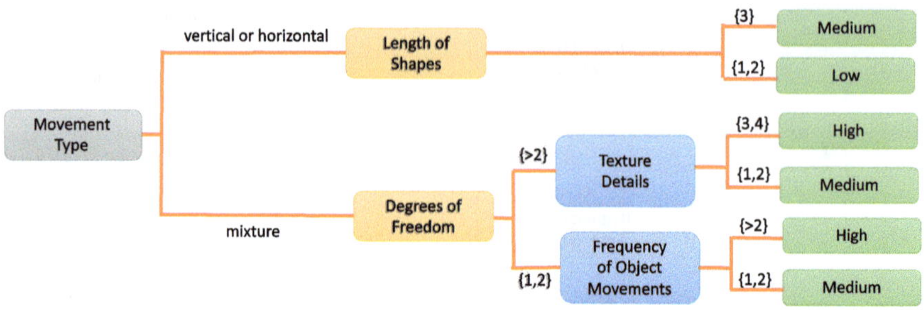

Fig. 3.12 Decision tree determining the encoding complexity of a game scene (cf. [83])

While the classification was promising and offering good performance for different tasks [30], it was developed under two limitations. First, the focus group was relatively small, which raises the question if there are more characteristics that are not considered. Second, the classification was developed based on an objective metric of recorded scenes, which may introduce errors depending on the metric's performance. Therefore, the second study was conducted similarly with a larger 2-steps focus group of 11 participants to identify and quantify the relevant game characteristics and using the actual subjective ratings instead of an objective metric. As a choice of the dataset, the video quality ratings of CGVDS and interactive datasets are used. The detail about the study and focus group study can be found in [83]. In the second study, five characteristics turned out to form the decision tree with an accuracy of 96% due to one miss-prediction for the low class. The final set of characteristics contributing to the decision tree is listed below. The decision tree obtained in the first attempt is illustrated in Fig. 3.12. Since the second study is conducted on a larger scale using the subjective ratings as ground truth, this classification is used to assign the video complexity class to the video games used in this book.

Movement Type (MV): Movement type is defined as the total number of camera directions. When considering a video as a 2D representation, this characteristic refers to the directions in which the video is changing. The directions can be vertical or horizontal movements, as well as a mixture of movements (e.g., diagonal).

Length of Shapes (LoF): The characteristic Length of Shapes describes the summed length of contours (shapes) of moving objects averaged over the time of the game scene. Movements of objects or within objects as well as in the background or environment should be considered.

Degrees of Freedom (DoF): Degrees of Freedom is defined as the freedom of camera movement. There can be up to six degrees of freedom due to three possible translations (back and forward, left and right, or up and down) as well as three rotations (vertical axis and height) of the camera.

Frequency of Object Movements (FOM): The amount of object movements—this also includes elements that are not controlled by the player such as background objects—is defined as the percentage of the total duration of the time frame, in which game objects are moving, to the overall duration of the game scene.

Texture Details (TD): Texture details refer to the graphical details in the game and depend on the number of used quads (polygonal shapes made of triangles). The more polygons used, the higher are the texture details. The game environment as well as other elements such as characters or obstacles should be considered.

3.6.2 Gaming Delay and Frameloss Sensitivity Classifications

Several studies showed that game design plays an important role in the sensitivity of games toward delay. One of the early works on the sensitivity of games to delay was conducted by Claypool, where the delay sensitivity was determined based on two factors, interaction and perspective [84]. Interaction is characterized by the deadline and precision model, which is discussed in Sect. 2.6. The game perspective specifies whether the game is based on an omnipresent model or an avatar model, in which the first-person and third-person perspectives are available. Claypool showed that the first-person avatar is significantly more sensitive to delay compared to the third-person avatar perspective in a shooting game, while the selected omnipresent game turned out to be less sensitive than avatar games [85].

In order to derive the game characteristics that influence the delay sensitivity of video games, similar to video complexity classification, a focus group methodology was conducted. In total, nine participants in three focus groups (each with three participants) are interviewed, which derived nine game characteristics that play a role in gaming QoE in the presence of delay. First, a short introduction was given to the participants before the interview in order to provide knowledge about the cloud gaming service and how delay affects gaming QoE. In addition, some well-known concepts of precision and deadline (cf. Sect. 2.6) are explained to the participants. Next, participants played 12 different scenarios picked from different game genres and game mechanics under different delays of 0 ms, 150 ms, and 300 ms. Participants played each selected scenario of games for 60 s, which are selected as a representative scenario of the game. The games are selected carefully to cover a wide range of game genres and game mechanics. After playing all selected scenarios, their observations on the characteristics that make a game to be sensitive towards delay are collected in a written text. Next, an open discussion led by a moderator took place in which participants in each focus group could discuss their opinion about the factors and characteristics that make a game sensitive to delay. By analyzing the interviews and notes, nine characteristics are selected as candidates to classify the games according to their sensitivity towards delay. The details of the study can be found in [81]. A very similar study was conducted for the frameloss

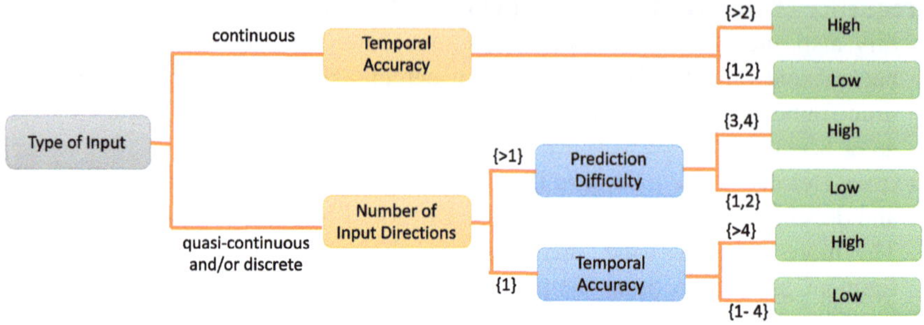

Fig. 3.13 Decision tree determining the delay sensitivity of a game scenario (cf. [81])

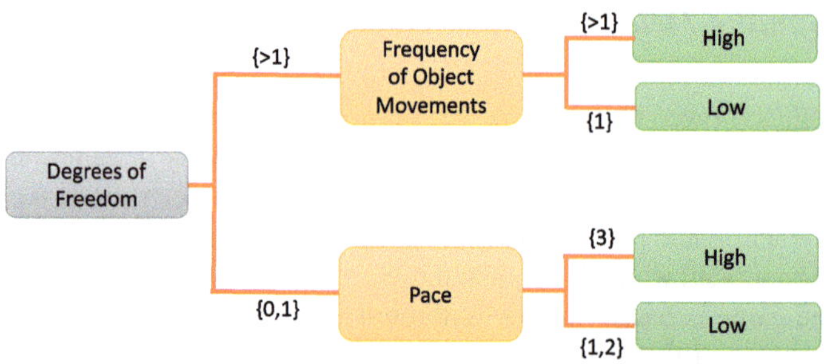

Fig. 3.14 Decision tree determining the frameloss sensitivity of a game scenario (cf. [83])

sensitivity to identify the relevant characteristics to classify the video games using three framerate levels of 10, 20, and 60 fps, that for the sack of brevity, it is not described here. Based on the input quality (GIPS items) ratings of a large dataset collected through a crowdsourcing approach [86] and the presented interactive dataset, two separate decision trees are built using the K-Means clustering algorithm to derive the frameloss and delay sensitivity classes. It turned out that DoF, FOM, and PC (described below) are the most important characteristics to classify the video games according to their sensitivity towards frameloss while, ToI, NID, TA, PR, NRA would be the necessary characteristics to build a decision tree for delay sensitivity classification. The decision trees for determining the delay sensitivity and frameloss sensitivity are presented in Figs. 3.13 and 3.14, respectively. The description of the characteristics can found below (note that DoF and FOM are already described in the previous section). The details of decision tree development and focus group study can be found in [81, 83].

(a) **Temporal Accuracy (TA):** Temporal Accuracy describes the available time interval for a player to perform the desired interaction. The time interval is strongly dependent on the mechanics and pace of a game scenario. In other words, this game characteristic describes the available reaction time of a player.

(b) **Predictability (PR):** Predictability describes if a player is able to estimate the upcoming events in the game. This can, for example, relate to positions of objects (spatial) or time points of events (temporal).
(c) **Pace (PC):** The pace of the game is defined as how fast the visible game elements (e.g., environment, characters, or obstacles) in the video are changed.
(d) **Number of Input Directions (NID):** The number of possible input directions in a game scenario is known as Degree of Freedom (DoF). DoF consists of translations (back and forward, left and right, up and down) as well as rotations (vertical axis and height) for one or multiple input devices/elements.
(e) **Number of Required Actions (NRA):** The number of required actions and with that also the number of inputs a player performs in a certain time frame may influence the perception of a network delay. The characteristic could also be described as the minimum actions per minute (APM) to play the game scenario. It is assumed that a higher number of required actions will lead to more user inputs and, thus, more situations in which a player can perceive a delay. The number of objects to react to or the pace of the game can also influence the number of required actions.
(f) **Type of Input (ToI):** The type of input describes the temporal aspects of player inputs on a spectrum of discrete to continuous. In some games, players are continuously giving input, for example, in a shooting game where players are always moving their mouse. Some games have discrete inputs meaning that players interact using pressing a button, for example, a jumping game where players must jump using pressing a key. In games with Quasi-Continuous inputs, players interact with the game using holding a key or constantly pressing a key.

According to the classification presented in this section, the games that are used to develop the presented interactive dataset are annotated to a different class of complexity/sensitivities in Table 3.5. Based on the table, it can be seen that a balanced distribution of games is selected for the interactive test except for the frameloss sensitivity class.

Table 3.5 Assigned class of complexity to games used in the interactive dataset

Game name	Delay sensitivity	Frameloss sensitivity	Encoding complexity
Bejeweled 3	Low	Low	Low
Counter strike	High	High	High
Dota II	High	High	Medium
Hearthstone	Low	Low	Low
Overwatch	High	High	High
Project cars	Low	High	Medium
Rayman legend	High	High	Low
Tekken	Low	High	High
Worms	Low	Low	Low

3.7 Summary

In this chapter, the necessary steps that are required for the development of a quality model for cloud gaming services are described. In a summary, the following items are described for the development of gaming QoE models:

- The scope and structure of the model are described. The model follows a modular structure, which allows the training of the model based on passive and interactive datasets separately. In addition, this modular structure could ease the future extension of the model for possible updates such as training for a new video codec by simply retraining the relevant impairment factor (e.g., I_{cod_V}) without changing other factors.
- Five image/video/interactive quality datasets are described in this chapter that can be used for training the gaming QoE model. The image and video quality datasets will be used for training the video coding impairment, while the interactive dataset allows training the core model and other impairment factors.
- The data post-processing method, including the data cleansing and outlier detection that are applied to each dataset, is explained. In addition, the transformation of scale from the assessed 7-point EC scale to a 100-point R-scale is described, which later will be used for training the impairment prediction models.
- Three video gaming classifications are proposed to classify games according to their sensitivity towards delay and frameloss, as well as video content complexity. The decision trees obtained a very high accuracy on the training dataset, above 0.90%. These classifications are great assets to develop an accurate gaming quality model.

While this chapter forms the basis of the development of quality models, the following chapters describe how to develop QoE models based on the model structure, different gaming quality datasets, and gaming classification presented in this chapter. In the next chapter, the models that are developed for the prediction of the video coding impairment are discussed.

Chapter 4
Video Coding Impairment Models

In total, the author contributed to the development of six models presented in this chapter and collaborate with other researchers to map ITU-T P.1204.3 predictions for gaming video datasets. Table 4.1 gives a short overview of the models discussed in this chapter and the datasets used for training and validation of the model. The models are trained based on different datasets, mostly presented in Sect. 3.4. The reason to use different training datasets for each model is either due to the time of model development, when some of the datasets were not yet created, or decisions in the design of the model that required to have access to diverse content or the high number of encoding parameters. "Training CGVDS" (referred in the table) includes approx 66% of CGVDS split based on the source video, 10 out 15 source sequences (together with all encoded sequences), and "Validation CGVDS" consists of the remaining video sequences (Fig. 4.1).

It has to be noted that the author did not contribute to the ITU-T Rec. P.1203 [87] and P.1204.3 [88], but because of the high performance of these models on gaming datasets, as well as minor correction to P.1204.3 for gaming videos, they are presented in this chapter.

Also, it has to be noted that in all calculations of this chapter, for simplicity of the results, the prediction is reported on a 5-point MOS scale, thus, they are referred to as video quality models. However, for the final gaming QoE prediction model, the predicted delta R-scale values are used.

This chapter is written based on the reference publications of video quality models that the author contributed to, presented in [50, 51, 89–91].

4.1 Planning Models

In this section, three planning models, also known as opinion prediction models, for video quality prediction, are presented. Planning models can be used by network

S. Zadtootaghaj, *Quality of Experience Modeling for Cloud Gaming Services*,
T-Labs Series in Telecommunication Services,
https://doi.org/10.1007/978-3-030-98249-2_4

Table 4.1 An overview of proposed video quality models, and datasets used for training and validating

Model	Model type	Training dataset	Validation dataset	Notes
ITU-T Rec. G.1072	Planning model	Training CGVDS, ITU-T Dataset	Validation CGVDS, KUGVD, GVSET	ITU-T Dataset is similar to CGVDS with different gaming content
GamingPara	Planning model	Training CGVDS	Validation CGVDS, KUGVD, GVSET	–
BQGV	Bitstream-based Model	CGVDS	CGVDS, KUGVD, GVSET	The model is trained and validated following one-leave-out cross validation
FHD P.1204.3	Bitstream-based Model	KUGVD, GVSET	CGVDS	The model is remapped based on each dataset
NR-GVQM	NR Signal-based	Part of GVSET	CGVDS, KUGVD, GVSET	For training the part of GVSET that was not a part of subjective test is used
NDNetGaming	NR Signal-based	Extracted frames of GVSET, GISET	CGVDS, KUGVD, GVSET	–
DEMI	NR Signal-based	Extracted frames of GVSET & NFLX-SVQD, GISET	CGVDS, KUGVD, GVSET	NFLX-SVQD is non-gaming dataset (cf. Sect. 4.3.3)

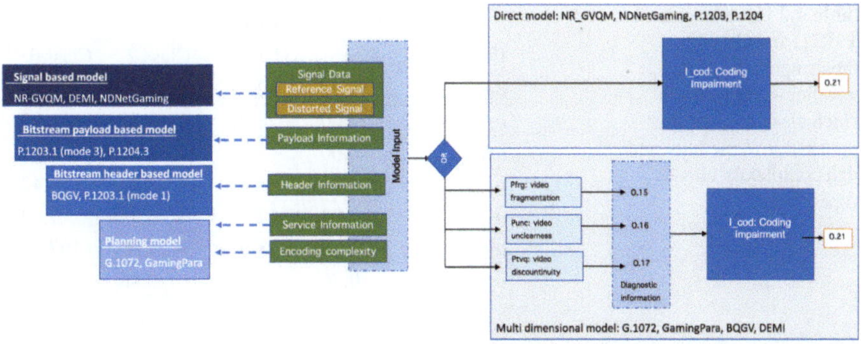

Fig. 4.1 Overview of objective video quality assessment models discussed in the book according to different levels of information extracted from the media stream

planners that have assumptions about the network, compression, and end-user characteristics. These characteristics could be network parameters, e.g., expected end-to-end delay, encoding parameters, e.g., expected encoded video bitrate, and device parameters, e.g., typical display size. Three models are presented, two ITU-T Recommendations where the author contributed to one of them, and another model that the author developed using the multidimensional approach. Firstly, ITU-T Rec. G.1072 and G.1071 are discussed. ITU-T G.1072 is an ITU-T Recommendation that predicts the gaming QoE based on multiple impairment factors. Thus, in this section, only the video coding impairment of ITU-T G.1072 is discussed (for more information, please refer to Sect. 5.4). Finally, GamingPara is proposed, which is developed following a multidimensional approach. Such a multidimensional approach benefits from the flexibility of the model for future updates and provides diagnostic information about the reasons behind the low-quality prediction. This section is written majorly based on the publication [50].

4.1.1 ITU-T Rec G.1071 and G.1072

The ITU-T Rec. G.1071 describes a planning model that contains a series of models for estimating the audio-visual quality of streaming services. The models are designed to be network planning tools. Therefore, they only use properties of the network and encoding setting as input parameters. Since the G.1071 models are not trained for gaming video streaming applications, the Annex A of the model is retrained for cloud gaming service based on the latest cloud gaming encoding setting (e.g. using hardware accelerator engines) in the development of ITU-T Rec. G.1072. Hence, both ITU-T Rec. G.1071 and G.1072 use the same model structure to predict video coding impairment but with different coefficients. The video impairment of G.1071 Annex A has six input parameters where three of them are set to zero, $packetloss = 0$, $slicesPerFrame = 0$ and $TSBurstiness = 0$, since, the

Table 4.2 Coefficients of G.1071, and G.1072 impairments for each encoding complexity in which class 1, 2 and 3 represent low, medium and high complexity classes respectively

	G.1071	G.1072		
Coefficient	–	Class 1	Class 2	Class 3
a_{1V}	51.28	52.51	37.99	47.75
a_{2V}	–22.00	–28.02	−13.72	−12.07
a_{3V}	6.00	−2.68	8.57	9.05
a_{4V}	6.21	5.47	3.27	3.42
a_{31}	3.92	12.42	6.83	7.62
a_{32}	–27.54	–28.019	−127.99	–167.84
a_{33}	0.26	0.22	0.48	0.076

transmission error in cloud gaming services is typically not introducing slicing effects but rather freezing artifacts (cf. Chap. 5). The remaining parameters that are used for model prediction are bitrate, number of bits per pixel ($BitsperPixel$), and framerate. The model predicts the impairments due to compression based on an exponential function of $BitsperPixel$ and linear relation of Content Complexity. $BitsperPixel$ is defined as the number of bits assigned to each pixel in average for a video sequence, which is calculated based on the number of pixels in each frame (the product of the width and height of video), bitrate, and framerate of the video. Content Complexity is a function of $BitsperPixel$. In order to predict the video quality of G.1071 Annex A, the core model follows the E-model impairment structure. Equations 4.1–4.4 show the model structure. The coefficients of the equations are presented in Table 4.2 for the G.1071 model and the refit G.1071 based on the gaming video dataset to drive coding impairment of G.1072. The G.1072 coefficients are given per class of video complexity (cf. Sect. 3.6) as can be seen in the table. As discussed earlier, the transmission impairment (I_{traV}) is set to zero.

$$VQ = 100 - I_{VQ-cod} - I_{traV} \tag{4.1}$$

where $I_{traV} = 0$ due to no transmission loss.

$$I_{VQ_{cod}} = a_{1V} \cdot exp(a_{2V} \cdot BitsperPixel) + a_{3V} \cdot ContentComplexity + a_{4V} \tag{4.2}$$

$$ContentComplexity = a_{31} \cdot exp(a_{32} \cdot BitsperPixel) + a_{33} \tag{4.3}$$

$$BitsperPixel = (Bitrate \cdot 10^6)/(NumPixelPerFrame \cdot Framerate) \tag{4.4}$$

4.1.2 GamingPara

In this section, a proposed planning model is described that can predict the overall gaming video quality based on the perceptual video quality dimensions, namely, fragmentation, unclearness, and discontinuity which are presented in Sect. 2.5.5.2.

Table 4.3 Coefficients of I_{VD} for each frame loss sensitivity class.

Coefficient	Low sensitive	High sensitive
d_1	66.29	67.28
d_2	10.23	3.23

Similar to ITU-T Rec. G.1071 and G.1072, the proposed planning model predicts the video quality solely based on the encoding parameters, bitrate, frame rate, and encoding resolution, but following a multidimensional-based approach. Such a structure-based model development is beneficial for a service provider as it is easier to update the model in case of new parameters or interest in a higher range of parameters. In addition, the multidimensional quality model can act as a diagnostic model to explain which degradation causes a potentially low video quality. The core model is developed based on impairment factors due to fragmentation, unclearness, and discontinuity, as shown in Eq. 4.5.

$$R_{QoE} = R_{(max,QoE)} - a \cdot I_{VD} - b \cdot I_{VF} - c \cdot I_{VU} \tag{4.5}$$

Each impairment factor is predicted solely based on parameters relevant to the representative dimension, which also offers a parameter reduction. The impairments and core model of the planning model are developed based on the "Training CGVDS" dataset described in the previous section.

Impairment of Video Discontinuity (I_{VD}) The video discontinuity impairment estimates the video jerkiness due to low encoding framerate. The impairment of video discontinuity is modeled based on an exponential function of frame rate as shown in Eq. 4.6. Two classes of temporal complexity are derived based on the video gaming classification presented in Sect. 3.6. If the network provider does not have any assumption about the class, the high complexity class should be used as a default mode (Table 4.3).

$$I_{VD} = exp(\frac{d_1}{framerate}) + d_2 \tag{4.6}$$

Impairment of Video Fragmentation (I_{VF}) The video fragmentation dimension represents the blockiness in the encoded video, which is mainly triggered by the encoding bitrate. The Impairment of Video fragmentation is modeled based on the bitrate, resolution, and framerate. Similar to G.1071, the variable $BitsperPixel$ is defined which explains the number of bits that is spent on average for a pixel, as shown in Eq. 4.7, where bitrate is defined in $kbps$ unit.

$$BitsperPixel = \frac{bitrate}{framerate \cdot height \cdot width} \tag{4.7}$$

Three encoding complexity classes are derived based on the gaming classification explained in Sect. 3.6. If no information about the class is available, the high

Table 4.4 Coefficients of I_{VF} for each video complexity class.

Coefficient	Low complex	Medium complex	High complex
e_1	21.1	13.79	11.21
e_2	−5.426	−8.017	−10.59
e_2	0.0005258	0.0005442	0.0006314

Table 4.5 Coefficients of I_{VU} for each video complexity class

Coefficient	Low complex	Medium complex	High complex
f_1	4.299	18.58	17.13
f_2	−2.016	−3.422	−4.494
f_2	−17.99	−15.38	−7.844

complexity class should be used as a default mode. I_{VF} is predicted based on Eq. 4.8 using the coefficients per class presented in Table 4.4.

$$I_{VF} = e_1 + e_2 \cdot log(bitperpixel \cdot bitrate) + e_3 \cdot bitrate \tag{4.8}$$

Impairment of Video Unclearness (I_{VU}) The video unclearness dimension represents the blurriness in the video caused due to upscaling the video resolution, e.g., if a bicubic or a bilinear interpolation is applied. Therefore, the scale ratio plays an important role, which is measured by the ratio of encoding resolution and display resolution, as shown in Eq. 4.9.

$$scaleratio = \frac{height_{encoding} \cdot width_{encoding}}{height_{display} \cdot width_{display}} \tag{4.9}$$

In addition to the scale ratio, the bitrate and amount of bits per pixel play an important role. Therefore, the impairment of video unclearness is predicted based on the two logarithmic functions of $BitsperPixel$ and bitrate as well as scale ratio, as illustrated in Eq. 4.10 using the coefficients presented in Table 4.5.

$$I_{VU} = f_1 + f_2 \cdot log(BitsperPixel \cdot bitrate) + f_3 \cdot log(scaleratio) \tag{4.10}$$

Core Model The core model of the proposed planning tool is based on the measured impairment as shown in Eq. 4.11.

$$R_{QoE} = R_{max,QoE} - 0.259 \cdot I_{VD} - 0.554 \cdot I_{VF} - 0.341 \cdot I_{VU} \tag{4.11}$$

The goodness of fit for the prediction of each dimension is reported in Table 4.6 in terms of RMSE and adjusted R-squared. It has to be noted that the RMSE is reported on the R-scale (ranges from 0 to 100).

Table 4.6 Goodness of fit for each perceptual dimension of GamingPara in terms of RMSE and Adjusted R-squared

	Class1		Class2		Class3	
	RMSE	Adj R2	RMSE	Adj R2	RMSE	Adj R2
Discontinuity model	10.43	0.41	6.96	0.60	–	–
Unclearness model	6.59	0.63	9.18	0.75	7.88	0.86
Fragmentation model	6.90	0.83	10.23	0.73	6.90	0.83

4.2 Bitstream-Based Models

In this section, three bitstream-based models are described. Two of these have been standardized as ITU-T Recommendations P.1203 and P.1204.3, which were developed within ITU-T Study Group 12/Question 14 (SG12/Q14) work item titled "P.NATS". While the author did not contribute to the development of the two ITU-T models, due to high performance pf P.1203.1, P.1204.3, and the FHD adapted version of P.1204.3 for gaming content, they are presented in this section. The author proposed a lightweight bitstream-based video quality model that is named BQGV. BQGV uses a multidimensional approach, which is shown to be a reliable approach for video quality prediction. Bitstream-based video quality models can be classified into the header-based and payload-based models, as explained in Chap. 2. While header-based models are significantly lighter, payload-based models, in contrast, are heavy in computation and typically are more accurate and less prone to the change of the coding setting due to access to detailed encoding settings, e.g., quantization parameters. ITU-T Rec. P.1203 has multiple modes covering both payload and header-based models, while ITU-T Rec. P.1204.3 is a payload-based model. The proposed BQGV uses only header information for quality prediction.

4.2.1 ITU-T P.1203

The ITU-T P.1203 is a series of models that are designed to predict the audio-visual quality for adaptive and progressive-download-type media streaming that could be placed in end-point locations or at mid-network monitoring points. ITU-T Rec. P.1203 has four modes depending on the level of access to the information due to encryption as shown in Table 4.7. Mode 0 of ITU-T Rec. P.1203 is a simple parametric-based model that predicts the video quality based on the encoding parameters. Equations 4.12 and 4.13 illustrate the functions that predict video quality (referred to as mos_{codV}) of Mode 0, where the $coding_{res}$ stands for video encoding resolution.

$$quant = a_1 + a_2 \cdot log(a_3 + log(bitrate) + log(bitrate \cdot \frac{bitrate}{coding_{res} \cdot framerate}) + a_4)) \quad (4.12)$$

Table 4.7 ITU-T Rec. P.1203 modes of operation

Mode	Encryption	Input
0	Encrypted payload headers	Meta-data
1	Encrypted media payload	Meta-data and frame size/type
2	No encryption	Meta-data and up-to 2% of the stream
3	No encryption	Meta-data and any information from the video stream

$$mos_{codv} = q_1 + q_2 \cdot exp(q_3 \cdot quant) \tag{4.13}$$

Due to change of coefficients for different modes, they are not provided in this section, please refer to the reference paper in [92].

Mode 1 is a bitstream-based model that uses the packet header information for quality prediction. For Mode 1, Eq. 4.12 is used for measurement of the *quant* parameter. However, the coefficients are updated, and the frame size was used instead of video bitrate. It must be noted that the content complexity is taken into account based on the average size ratio of the I-frames and none-I-frames. In addition, the mos_{codV} is predicted using a sigmoid function. Mode 2 can have access to 2% of the media stream. Therefore, the parameter *quant* is measured using Eq. 4.14 with higher accuracy using the quantization parameters of the bitstream:

$$quant = \frac{\overline{QP_{PB}}}{51} \tag{4.14}$$

where $\overline{QP_{PB}}$ is measured based on the quantization parameters in the bitstream for non-I-frames. The mos_{codV} can be predicted based on Eq. 4.13 for the Mode 2 model. The Mode 3 model has access to the full bitstream in which the QP values are parsed for all frames in the measurement window and uses the same equations as Mode 2. The main difference between Mode 2 and Mode 3 is the amount of information from the bitstream that can be accessed for training and predicting the video quality. In Mode 2, due to limited access to the bitstream, some QP values are estimated based on the neighboring frames, while in Mode 3 all QP values are used in training and prediction of the model. It has to be noted that the Mode 2 has been removed from the model implementation due to the practicality of usage.

4.2.2 ITU-T P.1204.3

ITU-T Rec. P.1204.3 describes a payload based bitstream-based video quality model developed within the P.NATS phase 2 project, which was conducted in collaboration between ITU-T and VQEG. The model targets two different devices, namely, TV/PC monitor and mobile/tablet (MO/TA) with 3840 × 2160 and 2560 × 1440, respectively as target resolutions, and multiple codecs (i.e., H.264, HEVC and Video Payload type 9 (VP9)), resolutions up to 4K/Ultra-High Definition-1 (UHD-1) and framerates up to 60 frames/s. The model aims at video streaming services over a reliable transport and covers multiple types of degradation that are common in cloud gaming services, such as blurriness and blockiness due to compression. The model defines three types of degradations that affect the video quality in a reliable transport protocol:

- *Quantization degradation* relates to the coding degradations that are introduced in videos based on the quantization settings which typically lead to blockiness artifacts. This degradation is predicted based on the average none I-frames, $QP_{-non-I-frames}$ and maximum QP (QP_{max}) which is set according to the video codec (e.g, for H.264-8bit, set at 51, and for VP9 and VP8 set at 255) as follows:

$$quant = \frac{QP_{-non-I-frames}}{QP_{max}} \tag{4.15}$$

$$D_{q-raw} = 100 - RfromMOS(a + b \cdot exp(c \cdot quant + d)) \tag{4.16}$$

- *Upscaling degradation* relates to the degradation introduced by upscaling a video sequence to the higher display resolution which leads to blurriness artifacts.

$$D_{u-raw} = x \cdot log(y \cdot \frac{coding_{res}}{display_{res}}) \tag{4.17}$$

- *Temporal degradation* relates to low framerate due to encoding or playing out the distorted video at a reduced framerate which leads to jerkiness artifact.

$$D_{u-raw} = z \cdot log(k \cdot \frac{coding_{framerate}}{60}) \tag{4.18}$$

For all raw determination models, the prediction is limited to the range of 0 to 100 (e.g., if D_{u-raw} predicts lower than 0, it will be reported as 0), to form to the final predictions as D_q, D_u, D_t.

Finally, the video quality is predicted based on the parametric-based model similar to E-Model as:

$$M_{parametric} = 100 - D_q - D_u - D_t \tag{4.19}$$

In order to improve the parametric-based prediction model, a Random Forest model is trained to estimate the residual predictions, i.e., the difference between the real video quality score obtained from subjective tests during model training and the prediction of the parametric part of the model. Thus, the final prediction model is linearly fit based on the parametric prediction and added residual from the Random Forest model as follow:

$$Q = w_1 \cdot M_{parametric} + w_2 \cdot M_{randomforest} \tag{4.20}$$

For brevity, the coefficients of each model are not presented in the book (please refer to ITU-T Rec. P.1204.3 [88]). Ramachandra Rao et al. [75] adapted the original model for the case of FHD as a display for gaming evaluation since the gaming databases that were evaluated had the target device as the screen with FHD resolution. As the P.1204.3 is trained based on TV/PC monitor with 3840 × 2160 resolution, in the prediction model for *Upscaling degradation*, the scale factor is slightly adjusted as follows:

$$D_{u-corr-fac} = a \cdot log(b \cdot \frac{coding_{res}}{1920 \cdot 1080}), \tag{4.21}$$

where $a = -0.1076$ and $b = 0.083$.

This adjustment is expected to minimize the over-predicting the *Upscaling degradation* for an FHD screen. Thus, the correction was added to the final prediction as follows:

$$pred_{hd-mapped} = pred_{p1204-3} + D_{u-corr-fac} \tag{4.22}$$

The model is then trained using the KUGVD and GVSET datasets and validated on the CGVDS dataset and another internal dataset made from Twitch gameplay recording. The performance of the adjusted model is reported in Chap. 6, where the FHD mapped version of P.1204.3 on gaming datasets is named as *FHD P.1204.3*.

4.2.3 Bitstream-Based Quality Prediction of Gaming Video (BQGV)

BQGV is a monitoring quality model that is developed using the bitstream features extracted from the packet header. Multiple bitstream features are extracted in a 5-second window duration which are listed below. The model does not require to extract the complex features such as quantization parameters, which allows the model to be categorized as a lightweight bitstream-based model, comparable to the ITU-T Rec. P.1203 Mode 1. To develop BQGV, multiple parameters are extracted from the packet header as listed below.

- fr: Encoding Framerate
- sr_{video}: scale ratio according to Eq. 4.9
- br_{avg}: Average bitrate of five second sequence
- $numI_{frame}$: Number of I_{frames}
- $bravg_I$: Average of I-frames bitrate
- $stat_{P-frame}$: Statistics of P-frames (average, standard deviation, etc.)
- $std_{P-frame}$: Standard deviation of P-frames bitrate

- CP_{video}: Video content complexity is taken into account based on the ratio of the I-frames and none-I-frames average size, $\frac{\overline{I_{bitrate}}}{P_{bitrate}}$, inspired by ITU-T Rec. P.1203.

Since the MOS ratings are post-memory judgments over the full duration—in the CGVDS 30 seconds—of a video stimulus, the quality of each 5-second clip of a video sequence is annotated based on the distribution of frame-level quality estimation of a gaming video quality metric, NDNetGaming (cf. Sect. 4.3), and final MOS values of each stimulus. The selection of NDNetGaming was due to the high accuracy of this model compared to others which is shown and discussed in Chap. 6. In order to annotate a 5-second interval of the video sequence according to MOS and NDNetGaming values, two steps are required. First, the quality weight of each frame is measured by the frame-level NDNetGaming score compared to the NDNetGaming prediction of the whole video sequence as shown in Eq. 4.23.

$$w_i = NDG_{i,vs}/NDG_{vs} \tag{4.23}$$

where, $NDG_{i,vs}$ is the NDNetGaming prediction of frame i in the video sequence of vs, while, NDG_{vs} is the predicted video quality of the video sequence vs, based on NDNetGaming prediction. Finally, the MOS of an interval duration from frame i to frame j ($\widehat{MOS_{i,j}}$) of the video sequence vs is estimated based on the weight of each frame contribution to overall video quality that can be measured according to Eq. 4.24.

$$\widehat{MOS_{i,j}} = MOS_{vs} \times \sum_{k=i}^{j} \frac{w_k}{j-i} \tag{4.24}$$

where MOS_{vs} is the Mean Opinion Score of the video sequence vs. The MOS for video quality, fragmentation, and unclearness is measured for the interval duration with the same approach.

Since all features are extracted over the 5-second duration, it is likely that an I-frame does not appear in the 5-second duration. CP_{video} estimates the spatio-temporal complexity of a video sequence based on the ratio of averaged I-frame and P-frame size in the measured window. Thus, CP_{video} is estimated slightly different if no I-frame exists in the selected 5-second duration. In that case, the I-frame from the previous window was used as a representative I-frame.

Similar to the proposed planning model, the BQGV follows the multidimensional approach that each video impairment is separately modeled, and then the video quality is predicted based on a prediction of impairments factors due to fragmentation, unclearness, and discontinuity. For training the model, a Random Forest (RF) regression method is used to train each impairment model. To reduce the model's complexity, a Recursive Feature Elimination (RFE) method is conducted to select the minimum number of features for each impairment factor.

Table 4.8 The performance of each single model to predict the impairment factors if full CGVDS is used as a training dataset in terms of RMSE, PLCC, and SRCC

	RMSE	PLCC	SRCC
I_{VU}	4.42	0.91	0.90
I_{VF}	5.21	0.92	0.89
I_{VD}	6.34	0.78	0.77

Impairment of Video Discontinuity (I_{VD}) Video discontinuity is the perception of jerkiness in the video due to the reduction of framerate by low encoding framerate. Among extracted features, the RFE analysis revealed the importance of three features that are also representative of temporal video dimension as follows: $fr, std_{P-frame}, CP_{video}$.

Impairment of Video Fragmentation (I_{VF}) Video fragmentation is affected mainly by the chosen bitrate but it can potentially be affected based on other extracted features. Therefore, the RFE method is applied to eliminate those that do not significantly contribute to the model. After applying RFE, the following relevant factors remained in the I_{VF} model: fr, $std_{P-frame}$, CP_{video}, sr, sr_{video}, $stat_{P-frame}$, br_{avg}.

Impairment of Video Unclearness (I_{VU}) Video unclearness is affected mainly by scaling ratio, while fragmentation in a video sequence might also affect the Video Unclearness. Therefore, all features are added to the model similar to the I_{VU}. Next, the RFE method is applied to select the most relevant features for the modeling. After applying RFE, six features remained in the model: fr, $std_{P-frame}$, CP_{video}, sr, sr_{video}, $stat_{P-frame}$.

The performance of regression models if all videos in CGVDS are used in training is presented in Table 4.8 in terms of RMSE, PLCC, and SRCC. It has to be noted that the RMSE is reported on the R-scale (ranges from 0 to 100). Based on the result, it can be concluded that the model could potentially reach high accuracy for the prediction of video quality.

Core Model The core model follows the impairment factor approach (cf. Chap. 3) with three impairments derived from the video quality dimensions. Equation 4.25 illustrates the core model structure and coefficients to predict the video quality.

$$R_{QoE} = R_{(max,QoE)} - 0.243 \cdot I_{VD} - 0.412 \cdot I_{VF} - 0.434 \cdot I_{VU} \tag{4.25}$$

To train and test the model fairly, the CGVDS dataset is split into five bins where in each bin, three games are randomly selected. The model is trained five times, each time one bin was held-out and the model is trained based on the remaining videos. In Chap. 6, the performance of the model on CGVDS is reported.

4.3 Signal-Based Models

In this section, the developed signal-based models for gaming content are presented. Previous research has shown a very low performance of existing NR metrics to predict the gaming video quality, as discussed in Chap. 2. NR methods require no knowledge from the original signal, which is challenging to reach a high accuracy in the prediction of gaming video quality because of the diverse spatial and temporal video characteristics of video games. Gaming content has special characteristics such as repetitive content, the same design style across a game, and a small game world, making this type of content a perfect target for training machine learning-based quality models as there is a high similarity in both temporal and spatial domains within the same game. Thus, the main focus of this section is to develop machine learning based NR models that learn the game features and characteristics in addition to the compression distortion for the quality prediction task.

Three models are proposed and discussed in this section. First, a lightweight NR model is described that uses the low-level image features (e.g., edges information) to predict the gaming video quality. Next, two deep learning-based models are proposed that are developed similarly but with differences in the model design and training process.

The models that are described in this section are written based on the text of three publications of [89–91].

4.3.1 NR-GVQM

NR-GVQM is a no-reference video quality model targeting gaming content and uses VMAF as the ground truth. NR-GVQM uses low-level image features for video quality prediction and reduces the computation time by keeping the number of required features to a minimum. The NR-GVQM metric is trained and validated based on the GVSET dataset (cf. Sect. 3.4).

4.3.1.1 Feature Extraction

NR-GVQM extracts a total of nine frame-level features from the videos. These include three image naturalness indices: Blind Image Quality Index (BIQI) [93], Natural Image Quality Evaluator (NIQE) [94] and Blind/referenceless Image Spatial Quality Evaluator (BRISQUE) [95]); two content complexity indices (Spatial Index and Temporal Index [3]), and four distortion specific indicators estimating levels of Noise, Blurriness, Blockiness, and Contrast based on the tool introduced in [96, 97]. The features were selected primarily based on three criteria: content complexity (SI, TI), existing NR metrics (BIQI, NIQE, and BRISQUE), which take into account distortions to the naturalness of an image/video and NR features which take into account distortions specific to the dataset used for training (measurement of Blockiness, Noise, Blurriness, and Contrast).

BIQI is a modular NR metric that tries to predict the image quality based on trained Natural Scene Statistic (NSS) under different types of image distortion. BRISQUE is also another image fidelity measurement metric that quantifies the possible loss of naturalness in an image by using the locally normalized luminance coefficients. Similarly, NIQE is a machine learning based index for evaluating the naturalness of an image. Although these three NR metrics are trained based on Natural Scene Statistics (NSS) features which might not be representative features for gaming content, they were found to have a satisfactory performance on gaming content (in line with their performance on non-gaming videos), especially at 1080p resolution (cf. Chap. 2). Alongside, measurements of Noise, Blurriness, Blockiness, and Contrast are used, as these distortions are more specific for the dataset under consideration. Additionally, per frame SI and TI values are used for training as they are often used as an indirect indicator for video content complexity.

4.3.1.2 Framework

In order to develop the new metric, the low-level features and VMAF values are extracted/measured for each frame of the training set, the GVSET. The GVSET dataset is divided into two parts. For the training set, 408 distorted video sequences that are not assessed in the subjective test (18 out of 24 reference videos) were used. 900 frames per video sequence (30 frames per second, 30 seconds total duration) and 408 distorted video sequences resulted in a total training sample size of 369000 frames for which nine features were extracted. The second part consisting of the remaining six reference videos with a total of 144 distorted sequences is used in the validation phase (the six considered reference videos are the same six videos as used for subjective assessment that is presented in Sect. 3.4).

As a choice of the machine learning method, a Support Vector Regression (SVR) is trained using linear and Gaussian Radial Basis Function (RBF) kernels. RBF is selected as the suitable kernel for SVR, based on some observations in relation to the features in the training dataset as well as to minimize RMSE values while maximizing $R{-}squared$. This procedure was carried out through trial and error. As a first glimpse on the performance of the obtained model, 10-fold cross-validation is applied in the training process, which gains an average RMSE of 3.56 for the training set. It needs to be noted that since the cross-validation was applied at frame-level, there is a high possibility that the different frames of the same video are used in training as well as in the validation process. Figure 4.2 shows the machine learning framework that is used for training and evaluating the model.

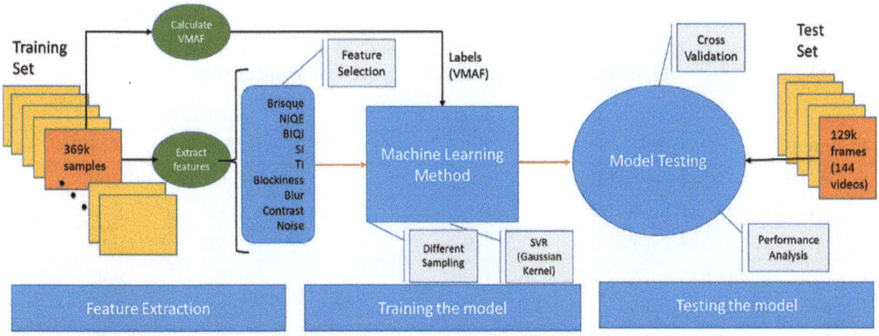

Fig. 4.2 Flowchart describing the NR-GVQM modeling approach

4.3.1.3 Computation Reduction

In order to reduce the computation cost, two approaches are examined. First the number of features is reduced by removing the features which do not significantly affect the performance. Additionally, the number of frames that are required per video sequence is investigated to decrease the computational cost while keeping the performance high.

4.3.1.4 Feature Reduction

Typically, for regression and classification problems, feature selection is an essential step before training any model in order to develop a better model for a given dataset. Feature selection can reduce the need for storage as well as computation. Besides, it can avoid overfitting and help in the interpretability of the model. Therefore, the features and their impact on the final model are analyzed. For feature selection, two approaches are examined, using domain knowledge as well as a feature selection technique in the machine learning domain. Based on the domain knowledge, it is expected to have a high correlation between the NR metrics used as features in the model, since they are all basically based on the same concept of NSS. Therefore, the model is trained multiple times with every possible combination of the image naturalness metrics and the other features. Moreover, it is investigated if SI and TI values have a significant impact on the final result. The second approach was using feature selection methods to choose the best set of features. Thus, a feature reduction method, the Neighbourhood Components Analysis (NCA) using Stochastic Gradient Descent (SGD) solver, was used. The result indicates the low importance of Contrast measurement as well as Noise. It has to be noted that NCA assumes that the relation between the features and labels is linear. This basis for feature reduction and it's outcome has to then be investigated if a model trained using all features except Contrast and Noise can still predict the VMAF with high performance or not. The results show that without Contrast and Noise, the model

Table 4.9 VMAF score prediction quality performance analysis in terms of PLCC and RMSE scores for different feature sets. Features are denoted as F1 to F9 corresponding to BIQI, BRISQUE, NIQE, SI, TI, Contrast, Blockiness, Blur and Noise respectively

	Gaussian Kernel		Linear Kernel	
Features	PLCC	RMSE	PLCC	RMSE
F1+F2+F3+F4+F5+F6+F7+F8+F9	0.98	4.73	0.88	8.81
F2+F3+F4+F5+F6+F7+F8+F9	0.97	5.14	0.88	8.84
F1+F3+F4+F5+F6+F7+F8+F9	0.97	5.28	0.88	9.08
F1+F2+F4+F5+F6+F7+F8+F9	0.96	6.29	0.88	9.47
F3+F4+F5+F6+F7+F8+F9	0.95	6.79	0.87	9.78
F1+F4+F5+F6+F7+F8+F9	0.97	5.88	0.88	8.89
F2+F4+F5+F6+F7+F8+F9	0.96	5.77	0.87	9.20
F1+F2+F3+F5+F6+F7+F8+F9	0.98	5.23	0.88	8.72
F1+F2+F3+F4+F6+F7+F8+F9	0.97	5.18	0.88	8.94
F1+F2+F3+F6+F7+F8+F9	0.93	7.66	0.85	9.36
F6+F7+F8+F9	0.89	9.95	0.72	12.79
F1+F2+F3+F4+F5+F7+F8	0.97	5.48	0.90	8.29
F3+F5+F7+F8	0.97	5.19	0.88	8.89

reaches a high performance of PLCC of 0.97 and RMSE of 5.48 w.r.t. VMAF on the training dataset. Table 4.9 illustrates the performance of the model using a different combination of features. It can be seen that using only four features (NIQE, TI, Blockiness, and Blur) can give a model with an accuracy of 0.97, which has a low computation but a high correlation with the actual VMAF values.

4.3.1.5 Temporal Sampling

Consecutive frames are very similar in terms of temporal and spatial scene characteristics. Therefore, to reduce the computation cost and time, the minimum number of frames that are required to reach a reasonable prediction is examined. The NR-GVQM scores are measured for a video sequence considering frames at specific intervals (e.g., every second or every third frame). Then, the performance behavior of the model in terms of PLCC scores and RMSE values for each considered interval is measured. Figure 4.3 shows the variation of RMSE value with an increase in the pooling interval. It can be observed that even though the correlation scores in terms of PLCC did not change much with the change in pooling interval, RMSE is very sensitive initially even for small changes in pooling (see Fig. 4.3). The scale for the PLCC curve is kept fine-grained to illustrate the drop in PLCC scores (local minimum in the figure) when the frame interval size is a coefficient of Group of Picture.

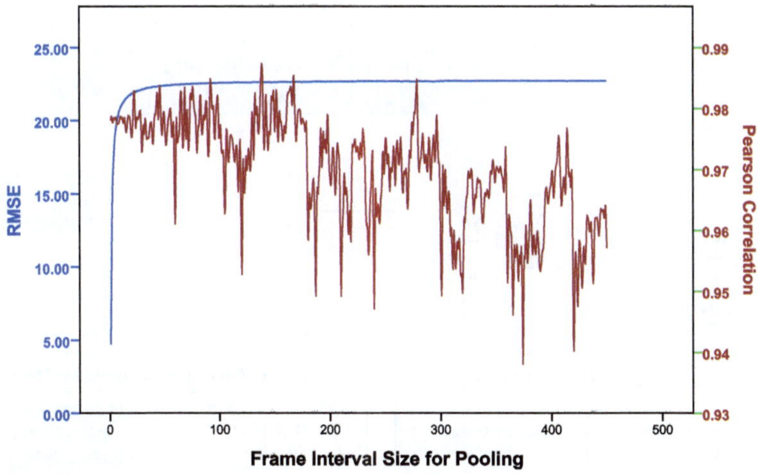

Fig. 4.3 Effect of different interval size for temporal pooling in terms of RMSE and PLCC scores

4.3.2 NDNetGaming

NDNetGaming is a deep learning based quality model that has two layers, a layer that predicts the image quality using a Convolutional Neural Network (CNN) and another layer to pool the frame-level prediction to predict the overall video quality based on the temporal aspects. In order to build a deep learning-based quality model, a large annotated quality dataset is required. Considering the fact that having access to such a large quality annotated dataset is not very realistic for any image or video quality task due to expensive subjective tests, it is even more crucial for gaming quality tasks due to a limited number of available datasets. Therefore, as an alternative way to overcome this limitation, a framework is designed in three phases. First, a CNN is trained based on an objective quality metric to allow CNN to learn different types of image degradation, e.g., blurriness and blockiness. Next, the model is fine-tuned based on a transfer learning method and by using a dataset of subjective image quality ratings, GISET. Finally, the video complexity estimation was used to pool the frame-level predictions to a video quality score. The three phases of the proposed framework are visualized in Fig. 4.4.

4.3.2.1 Fundamental Design Phase

To train the CNN, a frame-level prediction of the VMAF metric is used as ground truth for the image quality of all frames involved in this phase. With the aim of training and validating the frame-level model based on the VMAF values, the GVSET and KUVGD are used.

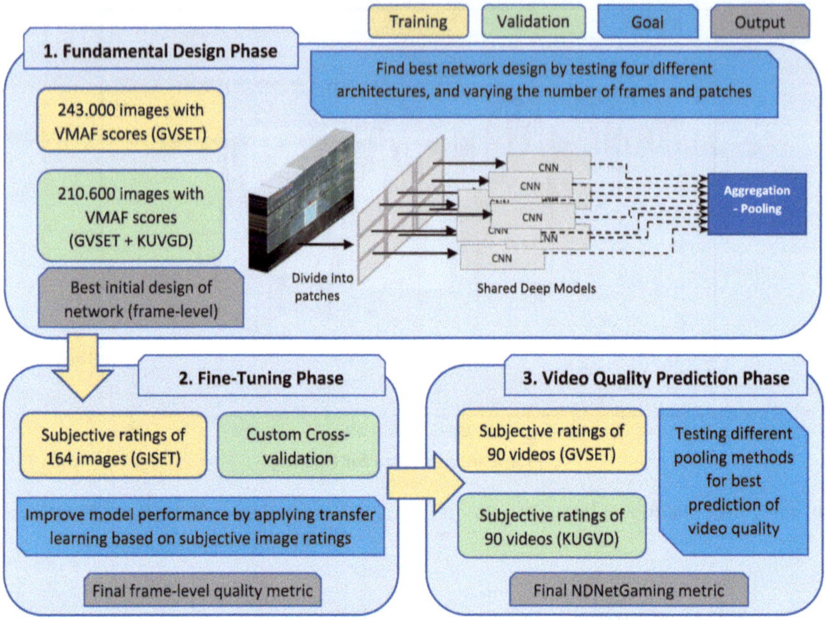

Fig. 4.4 Procedure of NDNetGaming metric development

All frames and corresponding VMAF values for 18 source video sequences (and their encoded sequences using 15 different resolution-bitrate pairs) from the GVSET were extracted and used in the training process of the model, for which no subjective video quality ratings are available. Thus, a total of 243.000 frames were used for the training of the CNN based on VMAF values.

Comparison of CNN Architectures

Using Keras [98], a high level neural network library, different model architectures are available along with their pre-trained weights on the ImageNet database [99]. Four popular architectures, DenseNet-121 [100], ResNet50 [101], Xception [102], and MobileNetV2 [103] are selected, and their performances on the training dataset are compared using VMAF labels. Each architecture has a different number of parameters and layers as well as different design styles to improve the information flow in the network.

Since the content of the training database is artificial, the similarity to the ImageNet content can vary a lot for different games. Therefore, it is difficult to decide how many layers of each CNN have to be retrained. For a fair comparison, the total number of trainable parameters of each architecture is reported and the results are compared when training only on the last 25%, 50% and 75% of parameters. The results are summarized in Table 4.10. Since the pre-trained networks are initially trained for a classification task, for each of these architectures the fully connected layer with multiple output neurons at the end of the network was removed in order

Table 4.10 RMSE on the GamingVideoSET for four different network architectures retraining only 25%, 50% or 75% of total trainable weights

	MobileNetV2	DenseNet-121	Xception	ResNet50
25%	9.59	7.58	7.33	7.60
50%	7.98	6.84	7.25	7.34
75%	7.34	6.74	7.29	6.71

to allow the model to get trained for the regression task. In addition, a dense layer consisting of only one output neuron with linear activation is added. The output of the network was directly compared to the actual VMAF values of the validation set that consists of different videos that were not used in the training session, and are encoded under various bit-resolution pairs settings. Because of the large size of the frames (1920 × 1080) in the dataset, random patches of size 299 × 299 from the frames were cropped for training the model. This was done in parallel to the training, such that in each epoch a new random patch of each image was chosen.

Based on the result presented in Table 4.10, it can be observed that ResNet50 and DenseNet-121 deliver the best results among the four architectures. Since DenseNet-121 has an advantage in terms of numbers of parameters over ResNet50, DenseNet-121 was selected for all following investigations. However, it should be noted that for an extensive evaluation, multiple hyperparameter settings should have been compared for every architecture. Also, for this comparison, every training was only done once per configuration, so the actual average numbers could vary slightly.

Required Number of Layers for Training

The pre-trained DenseNet-121 is trained based on a huge number of annotated images (over 14 million images) which is almost 100 times larger than our whole training dataset. In addition, in the early layers of a CNN, mostly basic features are learned by the models such as edges, shapes, corners and intensity which can be shared across different types of tasks. However, in the later layers of a CNN mostly features related to the task/application would be extracted and learned. Therefore, depending on the size of training data and diversity of content, there is an optimal point that the model is required to be retrained up to that point. In other words, training the whole DenseNet-121 model based on our dataset could lead the model to learn weaker in early layers due to lack of sufficient data or diversity of content.

To further investigate how many layers should be retrained for DenseNet-121, CNN is trained with differently sized parts of the network on the training dataset. The DenseNet-121 architecture consists of four blocks, each containing between 12 and 48 convolutional layers. Table 4.11 shows a comparison of the results after training of the model multiple times by increased step size of a half dense block in each training iteration. The results in terms of RMSE get better when more layers are used in the training process. However, this effect plateaus when it reaches 57 layers and even inverts when more than 107 layers are trained. The result in terms of SRCC does not vary much, except for the case where only the dense layer is trained.

Table 4.11 RMSE and SRCC for different choices of the number of convolutional layers

Dense blocks	Number of layers	Number of weights	RMSE	SRCC
4	120	7039k	8.11	0.925
$3\frac{1}{2}$	113	6878k	7.02	0.942
3	107	6657k	6.74	0.945
$2\frac{1}{2}$	94	6268k	6.77	0.946
2	82	5594k	6.84	0.942
$1\frac{1}{2}$	57	4461k	6.82	0.946
1	33	2191k	7.22	0.939
$\frac{1}{2}$	16	1233k	7.39	0.936
0	0	1k	10.60	0.870

Table 4.12 RMSE and R-squared for different interval size between two selected consecutive frames in the dataset

n-th frame interval	RMSE	R-squared
3	6.54	0.87
7	6.65	0.88
13	6.39	0.90
27	6.67	0.88
53	6.95	0.88
103	6.90	0.88
203	6.94	0.87
403	7.02	0.88

Patch Selection for Training

Since most of the common architectures have a fixed input patch size which is typically much smaller than the actual image size, it raises the question of how to select the patches in the training and testing process of the model. Therefore, it was investigated if only cropping the center of images would be a better choice for training the model compared to randomized cropping during the training of the CNN, given the fact that users tend to look at the center of images [104]. The results revealed that center crops cannot improve the performance of training compared to single random crops while taking multiple random patches from each frame in the training phase achieves higher performance.

Furthermore, the model is trained based on a smaller number of frames. Since the training dataset consists of over 300,000 frames and consecutive frames can be very similar, only every n-th frame from every video is used for training. In order to find a suitable step size n, a very high number was chosen ($n = 400$) and then lowered step by step to find the point, where the performance of the model stops improving. The analysis showed that $n = 13$ is the ideal threshold for our dataset in which lowering down the step size would not improve the result. The results are summarized in Table 4.12.

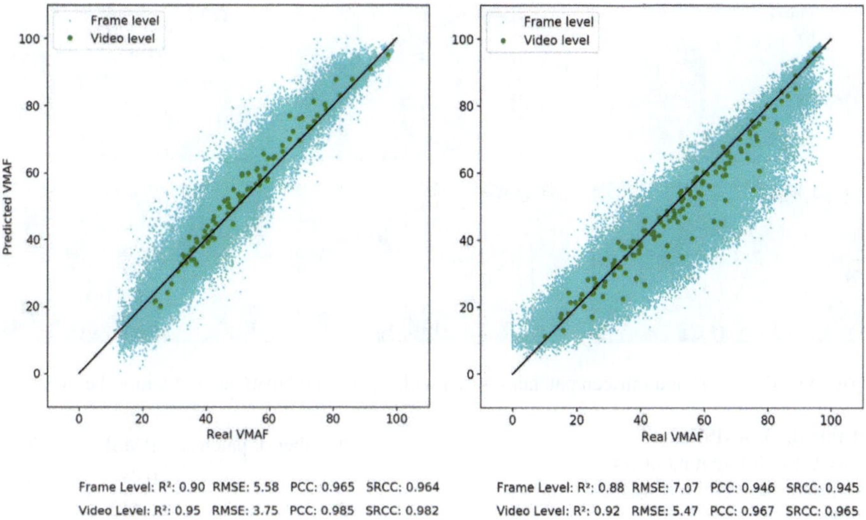

Fig. 4.5 Scatter plot of actual VMAF and predicted VMAF values on frame and video level of GVSET (left) and KUGVD (right) datasets

The Final Model for VMAF Prediction

To build the model that can predict the VMAF values on the image level, DenseNet-121 was trained after replacing the fully connected layer with a dense layer, taking 57 layers for training while using every 13th frame. The model reaches an RMSE of 6.19 with a PLCC of 0.954 on frame level and an RMSE of 2.62 and PLCC of 0.96 on the video level, using average pooling. Figure 4.5 shows the scatter plot of predicted and calculated VMAF values on the frame-level and averaged pooled video-level, for the unknown part of GVSET and KUGVD datasets (validation dataset).

4.3.2.2 Fine-Tuning Phase

As the last step to obtain an image quality model for gaming content transfer learning is used on the model that was already trained on VMAF values to further fine-tune the model using the subjective image quality ratings. Since the number of MOS-labeled images is limited, 13 patches are taken from each image, with a special pattern shown in Fig. 4.6 top right side, in the training phase. It has to be noted that the Differential MOS (DMOS) was used for the training which is measured according to ITU-T Rec. P.913 [105]. Due to the limited number of frames for the fine-tuning step, multiple patches from a frame are taken to improve the learning process. Specific patterns are designed for different numbers of patches to be taken from a frame to get a better representation of the image than by just taking patches randomly. Figure 4.6 shows some example patterns and compares them to a random

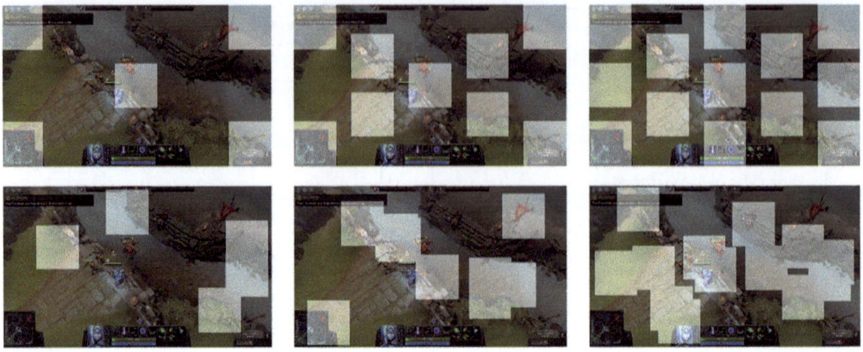

Fig. 4.6 Five, nine and thirteen patches chosen with a pattern (above) and randomly (below)

Table 4.13 RMSE and SRCC for different numbers of patches used for testing the model

Number of patches	RMSE	SRCC
1	0.481	0.894
3	0.413	0.944
5	0.390	0.953
7	0.374	0.957
9	0.380	0.954
11	0.381	0.958
13	0.377	0.953

patch choice. Since there are often special elements like texts or maps in the corners of a gaming video that are important for most of the games, the four corners are included in all patterns. Additionally, the center patch will be used in all patterns since participants in image tests tend to look at the center of an image [104].

Table 4.13 compares the results for different numbers of patches using the patterns that are designed. The results for multiple patches are higher than with only one patch per frame. However, it is difficult to conclude on the optimum number of patches because there is no clear trend. This is probably due to the small size of the image data set, GISET.

Due to the small set of images in the training set, a leave-one-out cross-validation was employed where for every iteration of training the network, one game is completely held out of the training process (reference video together with all encoded videos of the holdout game) and the model is evaluated based on the holdout game. This process repeats twelve times for every game in the GISET. Based on the custom cross-validation we obtained an aggregated (after testing all games) RMSE of 0.354 and PLCC of 0.959.

4.3.2.3 Video Quality Prediction Phase

Predicting video quality based on an image quality metric is typically done by average pooling of the frame-level predictions. While several pooling methods have been studied for gaming content [61], and non-gaming content [106], no significant improvement compared to the average pooling method has been observed. However, it has been shown that participants tend to rate the image quality worse than video quality due to cognitive load as well as temporal masking effect [61, 107]. Temporal masking is one of the important aspects of the human visual system (HVS), which has proven to have an impact on the perception of video artifacts. Choi et al. [108] analyzed the influence of motion on the performance of image level quality metrics after dividing a video quality dataset (LIVE VQA database) into two subsets of low-level motion and high-level motion contents. Their results revealed that many frame-based quality metrics such as PSNR, perform poorly in case of high-level motion content. A similar trend was observed in the training process of NDNetGaming where there exists a trend that the averaged pooled NDNetGaming prediction underestimates the quality of video with high-level motion content while overestimating the quality of content with a low level of motion. The performance of NDNetGaming using only average pooling is presented in Chap. 6.

Due to the temporal masking effect, the widely used average pooling method would not lead to achieving the highest video quality prediction, especially for content with high temporal complexity. Therefore, the image level predictions are pooled based on the temporal complexity of the video. Thus, higher weights are assigned for the contents with lower temporal complexity considering the temporal masking effect.

In order to measure the temporal complexity, the motion or optical flow pattern might be a good choice. However, motion estimation could get affected significantly by compression distortion such as blur and blockiness. Therefore, a simple measurement of temporal complexity (TC) is used based on the difference between consequent frames, similar to TI [109] but with a minor modification.

The per-frame TI values are calculated and the extreme values are ignored by using the exponentially weighted moving average (EWMA). Then the Inverse Probability weighting (IP weighting) of EWMA results is measured that gives a higher probability to small values compared to high values. The Inverse Probability weighting will be used to give higher weights to the low motion part of the game compared to high motion frames.

$$ewma_{TI} = smooth_{ewma}(std_{space}[M_n(i, j)]) \tag{4.26}$$

$$weights_{frame} = ewma_{TI}/sum_{time}[ewma_{TI}] \tag{4.27}$$

$$inverse_{weights} = \frac{(1 - P(F = 1))}{1 - P(F = 1|W = w)} \tag{4.28}$$

Table 4.14 Coefficients of temporal pooling methods, eq_{GVSET}, eq_{KUGVD} and $eq_{GVSET-KUGVD}$ are trained on GVSET, KUGVD and datasets combined, respectively

	c_1	c_2	c_3	c_4	c_5
eq_{GVSET}	−1.99	1.097	0.00069	−0.031	0.43
eq_{KUGVD}	−0.532	1.116	0.00011	−0.0043	0.084
$eq_{GVSET-KUGVD}$	−1.71	1.107	0.00053	−0.024	0.353

where $M_n(i, j)$ is the difference of pixel values between two adjacent frames considering only the luminance plane, and $smooth_{ewma}$ is the exponentially weighted moving average function. The weighting average is useful to pool the quality with more weight on low temporal complex frames. However, these weights are only considering the local temporal complexity of frames in a video and not between video sequences. In order to take into account the difference between the temporal activity of video games, the TC values for each video are measured to weight the quality ratings of complex videos in terms of temporal activities over the low complex videos. Therefore, a polynomial model is fitted to predict MOS values based on the average TC value of videos and frame-level NDNetGaming Score. The selection of the polynomial model was based on the observation of the relation between TC values and residual of NDNetGaming prediction and MOS values. Due to limited available data, the model is fitted based on the GVSET and tested it on the KUGVD and vice versa, for which the results are reported in Sect. 6.2. In addition to the two fitted models, a third model is built which is fitted on both gaming datasets for future works. Equation 4.30 presents the structure of the temporal pooling model, and Table 4.14 presents the coefficient of the model trained on different datasets.

$$TC = mean_{time}[std_{space}[M_n(i, j)]] \tag{4.29}$$

$$NDNG_{Temporal} = c_1 + c_2 \cdot NDNG + c_3 \cdot TC^3 + c_4 \cdot TC^2 + c_5 \cdot TC \tag{4.30}$$

4.3.3 DEMI

DEMI is a deep learning based video quality model that is designed similar to NDNetGaming with a few differences in the design phase. DEMI does target not only the gaming content but also non-gaming content to close the gap between metrics that only work for gaming content or non-gaming content. Therefore, DEMI trained based on both gaming and non-gaming datasets. Similar to NDNetGaming, DEMI uses DenseNet-121, and it has three phases of VMAF training, fine-tuning by subjective rating, and pooling the frame-level prediction to predict the video quality.

While NDNetGaming showed a high performance with gaming datasets, it has a few drawbacks that DEMI aims to address, as summarized below:

- NDNetGaming uses the frame-level quality as a representative quality label for each patch that is taken from the corresponding frame. However, it is known that not every patch in an image has the same level of quality.
- NDNetGaming predicts the quality for a fixed patch size that is projected only on 1080p resolution. This leads to difficulties to pool multiple patch quality predictions of one image into a single image quality score, where NDNetGaming simply uses the average pooling to predict the image level quality.
- NDNetGaming is biased to gaming content, and while it provides a high-quality prediction for gaming content, it does not perform similarly well for non-gaming content.
- NDNetGaming uses a very simple method to pool frame-level quality prediction to video quality score based on the only framerate of 30 fps. This leads to the inaccurate prediction for videos with a different level of framerate, which is discussed in Chap. 6.

The DEMI is proposed to address the drawback mentioned earlier, and also provides diagnostic information about the level of blockiness and blurriness in a video.

4.3.3.1 DEMI Model Architecture

In this section, the architecture of the proposed model, DEMI, and the special model design are described. DEMI is a CNN based metric that takes into account different types of artifacts such as blockiness, blurriness, and jerkiness, to predict the overall gaming video quality. The structure of the model is shown in Fig. 4.7. DEMI has three components. The first component is a CNN which is used to predict the frame-level fragmentation and unclearness. The second component is a temporal complexity indices which are based on Block Motion estimation (BM) and TI. Finally, the predicted fragmentation, unclearness, TI, and BM for multiple frames of a video are fused using a random forest model to predict the video quality which is the third component of the proposed model.

4.3.3.2 Phase 1—VMAF Training

In order to train a CNN for the image quality estimation task, a major limitation is the availability of a large scale image quality dataset with images and their subjective ratings. Mixing multiple datasets could be one option but it suffers strongly from many shortcomings such as subjective bias, the difference in viewing conditions, display used, etc., and hence, requires an anchor dataset to deal with this bias which is missing in such cases. Similar to NDNetGaming, the frames are annotated using VMAF values. As discussed in the previous section, training the

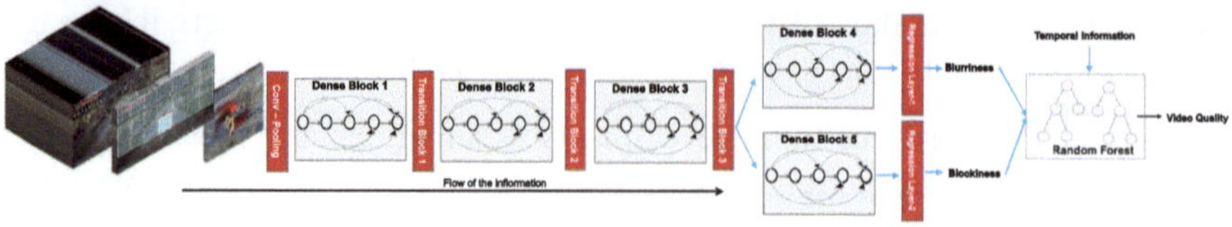

Fig. 4.7 Architecture of the proposed model (adapted based on [100]). Each transition block consists of 1x1 Conv and 2x2 Pool with stride 2. The regression layer has an average pool and a dense layer consisting of only one output with linear activation. Temporal information consists of TI and BM features

CNN using the VMAF annotated frames allows the network to learn different types of image compression degradation such as blurriness and blockiness.

As the underlying CNN architecture, DenseNET-121 architecture [100] is selected, as it was shown in NDNetGaming development that it performs well for image quality estimation tasks [91].

In its entirety, over 200k frames and their respective VMAF scores are used for training the model. The frames are extracted from multiple videos from several datasets. Nine patches are extracted per frame which makes the total number of patches during the training phase over a million. For training the model, two sizes of the image are used, full size (1080 × 1920), and half-size (540 × 960). This has been done to take into account the multi-scaling attribute in the training phase.

One of the major problems of NDNetGaming was to assign the quality of a frame (in terms of VMAF or MOS values) to the corresponding patches which distributes the error in the model since the quality of a patch is not necessarily the same as the quality of the corresponding frame. Thus, DEMI uses Partial PSNR to determine the quality and the weight of each patch that contributes to the overall VMAF score. Thus, for the patch i of frame j, the weight of the patch is calculated as follows:

$$W_{(i,j)} = PPSNR_{(i,j)}/PSNR_j \tag{4.31}$$

The quality of patch i from frame j is measured based on the VMAF of frame j which is calculated as:

$$VMAF_{(i,j)} = VMAF_j * W_{(i,j)} \tag{4.32}$$

The selection of PSNR is due to the simplicity and nature of the metric as it only measures the signal to noise ratio and avoids any content bias or scaling adjustment. Due to the high similarity between neighboring frames, in the training process, only every 20th frame is used for training.

4.3.3.3 Phase 2- Fine-Tuning

Once the model is trained based on VMAF, it is then fine-tuned two times based on a small image quality dataset using fragmentation (represents blockiness) and unclearness (represents blurriness) subjective ratings. 33 layers (i.e. last DenseNet Block) of the pretrained DenseNet-121, based on VMAF, were retrained using transfer learning. Approximately 25% of the CNN was retrained once based on the unclearness ratings and once based on the fragmentation ratings for the GISET dataset. Since only 25% of the CNN was retrained two times, the overhead of double training (for unclearness and fragmentation dimensions) does only result in computational overhead for testing the model for one additional DenseNet block due to the forward propagation of the model prediction phase.

4.3.3.4 Phase 3: Video Level

Once the model is fine-tuned based on the unclearness and fragmentation ratings, the frame-level prediction of the model is collected to be used in the training process at the video level. In addition to the frame level prediction, temporal features (temporal index and block motion estimation) are extracted for better prediction of the video quality. Random Forest (RF) was used as the training algorithm to fuse the features for the prediction of video quality.

Temporal Information (TI) between two consequence frames is defined similarly to ITU-T Rec. P.910 [3] as:

$$TI = std[M_p^n] \tag{4.33}$$

where M_p^n is the pixel intensity difference between F_p^n, current frame n and F_p^{n-1}, previous frame $n-1$, calculated as

$$M_p^n = F_p^n - F_p^{n-1} \tag{4.34}$$

Block Motion (BM) estimation with a block size of 8x8 is calculated based on Scikit-learn video library.[1] The block motion is then averaged over a frame (between two frames) and a single value per frame (second frame in each prediction) is stored for training. With consideration of the low computation complexity during the test (considering real-time prediction requirement in real-world applications), the frame-level features are extracted for every 20th frame.

[1] https://scikit-learn.org/stable.

4.3.4 Model Training

In this section, the training for each phase as well as the performance of each training phase in terms of PLCC, SRCC and RMSE are reported. For the training, the scale of VMAF was from 1 to 100 and for Phase-2 and 3, a 5-point ACR scale is used and RMSE is reported accordingly. For training phase, three datasets of GVSET, KUGVD and LIVE-NFLX-II Subjective Video QoE Database (NFLX-SVQD) are used. NFLX-SVQD [110] consists of 15 source videos and a total of 420 distorted sequences obtained by encoding the source videos at different bitrates at native resolution. The dataset includes both objective and subjective quality ratings, both continuous as well as retrospective prediction scores.

For validating the performance in each step of training, the CGVDS dataset and Netflix Public Dataset (NFLX-PD) are used. NFLX-PD is a non-gaming video dataset provided by Netflix [111] consisting of nine source video sequences of 1920×1080 resolution with framerates of 24, 25, and 30 fps. The videos are encoded in multiple resolution-bitrate pairs with bitrates ranging from 375 kbps to 5800 kbps and resolution ranging from 288p to 1080p.

Since for the subjective tests of the four datasets, the display resolution was set to 1080p, all videos are rescaled to 1080p resolution before extracting the frames. In the following, the detail on the training of each phase is given.

1. *Phase-1 (VMAF Training)*: In the first phase, the model is trained using VMAF scores from three datasets, GVSET, KUGVD, NFLX-PD. The DeneseNet-121 was trained based on the frames extracted from these three datasets, using the VMAF scores as the target labels. The result on the training set showed high performance with RMSE of 5.15 and PLCC score of 0.943 at frame level and RMSE of 3.25 and PLCC of 0.954 at video level (using average pooling) across all training datasets. The result of the two validation datasets is shown in Fig. 4.8.
2. *Phase-2 (Fine-tuning)*: Once the model is trained based on VMAF scores, it is then fine-tuned based on MOS scores from GISET, as it includes scores for both

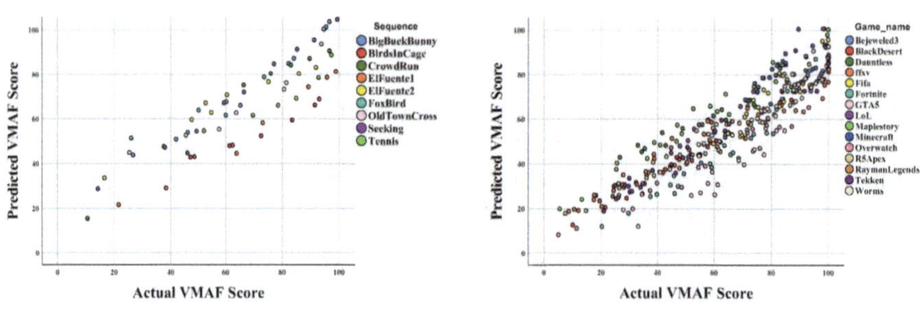

Fig. 4.8 Scatter plots of predicted VMAF vs actual VMAF scores for the two test datasets. (**a**) Predicted VMAF vs. actual VMAF scores for NFLX-PD dataset. (**b**) Predicted VMAF vs. actual VMAF scores for CGVDS dataset

fragmentation and unclearness. The model is fine-tuned in two steps, once using the scores for fragmentation and once based on scores for unclearness. The same weighting method, based on partial PSNR, explained in Phase-1 was applied to the ratings of each patch. Since the number of images is quite less, a leave one out cross-validation method was deployed where all video sequences from a game (reference video together with all encoded videos of that video sequence) are left out. The process is repeated twelve times for each game in the GISET. The result shows the high performance of the model for both blockiness, with PLCC of 0.94 and RMSE of 0.39, and blurriness with PLCC of 0.92 and RMSE of 0.45. Due to the small size of the dataset, all possible non-overlapping patches for fine-tuning the model are extracted.

3. *Phase-3 (Video-Level)*: In Phase-3, the model is trained at video level using four datasets, GVSET, KUGVD, NFLX-SVQD, and a subset of CGVDS consisting of videos of 60 fps (since the other datasets were limited to videos of up to 30 fps). A random forest model is used for training based on temporal features and the predicted fragmentation and unclearness predictions. The features are extracted only from nine patches of a frame and only from every 20th frame. The statistical information (average and standard deviation) of patch features over a video is also used in training the random forest. The result for the training data showed a very high PLCC score of 0.941 and RMSE of 0.31. The performance on the validation datasets is presented in Chap. 6.

4.4 Summary

In this section, multiple video quality models are proposed to predict the video quality of gaming content. The models can predict the quality in terms of the 5-point MOS scale or the R-scale that will be used for integration of the coding impairment into the gaming QoE prediction model. The proposed models are designed for different purposes as summarized below:

- Planning models can be used by network and service planners for better resource allocation. They can also be used as simple parametric-based models to predict the video quality based on simple encoding parameters. The author of the book developed the GamingPara model that follows the multidimensional approach. In addition, the author contributed to the development of ITU-T Rec. G.1072 that is presented in this chapter.
- A bitstream-based model is proposed for monitoring purposes, named BQGV. The model can be employed in both end-point locations and at mid-network monitoring points, in order to monitor the video quality of cloud gaming services. The bitstream information can be extracted using network probes if the bitstream is not fully encrypted. Some cloud gaming providers, e.g., NVIDIA Geforce Now, are already using this information to monitor the quality. The BQGV is developed because of the low performance of existing bitstream-based models

that only use packet header information on gaming video quality datasets which is discussed in Chap. 6. Also, two ITU-T Recommendation models are described as a candidate for quality assessment of gaming video streaming applications.

- Three NR signal-based models are described in this section. The FR models are not considered in this work since it has been shown in multiple publications [50, 60, 61, 63] that the current SoA FR video quality models perform well with gaming content, while NR models fail in high video quality prediction. While one of the proposed models is a lightweight video quality model, the two other models are deep learning based models for accurate quality prediction. These models can be used for multiple purposes. First, they can predict the quality of recorded gameplays for passive video streaming platforms, such as Twitch, which the reference signal is not available. Second, the NR metrics could be used for comparison of different cloud gaming services and their encoding strategies. Finally, the NR metrics can be used for cloud gaming service quality prediction without recording the reference signal to avoid heavy processes to servers and waste storage.

In the next section, the integration of coding impairment models presented in this section with other components of the gaming QoE model is described.

Chapter 5
Integration of Impairment Factors to Gaming QoE

5.1 Input Quality

Input quality is an important quality aspect that can directly impact the interaction of players with the game. As discussed in Sect. 2.5.5, controllability and responsiveness are identified quality features of the input quality aspect. Previous research showed that with respect to network and encoding parameters, input quality in the context of cloud gaming services could be affected by delay, packet loss, and low encoding framerates that occur during the gameplay.

Delay in a cloud gaming service is introduced by multiple processes such as the packet propagation and transmission, game tick rate, scene rendering, video and control packet encoding/decoding, as well as the refresh rate of the display and processing of the input device on the client-side. However, in all calculations presented in this chapter, when referring to a delay or latency, the round-trip time for transmitting a video and control command stream (sum of transmission and propagation delay) is considered, excluding the above mentioned delays due to other processes between client and server. Although all types of delay are important and play an important role in the interaction of players, from a practical point of view, it is not possible to accurately measure them at network monitoring points. In addition, with the advancement of encoding technologies, the largest proportion of delay in a cloud gaming service results from network delay consisting of transmission and propagation delay. Propagation delay refers to the amount of time that a signal takes to traverse the medium, e.g., a metal wire. Transmission delay refers to the transmission rate of an interface, e.g., a router in the network.

Furthermore, packet loss can affect the video stream either by the presence of slicing or freezing artifacts. Slicing artifacts are introduced when packet losses are concealed through the use of a Packet Loss Concealment (PLC) scheme to repair erroneous frames [112]. Freezing artifacts are introduced when the PLC scheme of the receiver replaces the erroneous frames (either due to packet loss or error

S. Zadtootaghaj, *Quality of Experience Modeling for Cloud Gaming Services*,
T-Labs Series in Telecommunication Services,
https://doi.org/10.1007/978-3-030-98249-2_5

propagation) with the previous error-free frame until a decoded image without errors has been received. Since the erroneous frames are not displayed, this type of artifact is also referred to as freezing [112]. Typically, cloud gaming services conceal the effect of packet loss by a freezing method due to fast concealment, which is more convenient for real-time streaming services. However, there is no common packet loss concealment method shared among cloud gaming services. Depending on the cloud gaming service, the packet loss concealment strategy might result in different levels of freezing received on the client-side. For example, some cloud gaming services, e.g., Google's Stadia, reduce the bitrate in case of a high packet loss rate to lower the frequency of freezing artifacts. Therefore, in order to not develop a model that is strongly service dependent, the effect of packet loss on the input quality is measured based on a parameter named *average frames per second* (AVG_{fps}) that can be measured and updated for any cloud gaming services by running multiple simulations.

AVG_{fps} is the average number of successfully received frames per second in a certain period of time. AVG_{fps} can be affected based on packet loss, bitrate, framerate, and delay. If a high level of bitrate is selected to encode a video sequence, more packets are required to stream the video, which increases the chance that more frames are affected by packet loss compared to the case when a low bitrate is selected. In addition, if the delay is very low, the missing packet can be re-transmitted, which reduces the number of lost frames. However, re-transmission introduces an additional delay, which might lead to lower gaming QoE. Therefore, depending on the latency of service, game content, and level of packet loss, cloud gaming providers might adopt different strategies to conceal packet losses.

For the cloud gaming service that is used in the interactive dataset, the AVG_{fps} is measured by recording several traces of simulated network and encoding settings (under different packet loss, bitrate, framerate, and delay levels). Based on the collected data through several simulations, the AVG_{fps} is measured for the selected cloud gaming service in the interactive dataset as shown in Eq. 5.1.

$$\text{Avg}_{fps} = FR_{enc} \cdot exp((d_1 + d_2 \cdot FR_{enc} + d_3 \cdot Bitrate \cdot FR_{enc}) \cdot (d_4 \cdot \frac{Delay}{25} - d_5) \cdot PL) \quad (5.1)$$

where, $d_1 = 0.08526$, $d_2 = 0.00073$, $d_3 = 1.425e - 04$, $d_4 = 2.414$, $d_5 = 1.5$, and FR_{enc} is the encoding framerate.

For a new cloud gaming service, the AVG_{fps} should be modeled again similarly using several simulations. Based on the measured AVG_{fps}, the $FrameLossRate$ can be calculated according to the AVG_{fps} and FR_{enc} as shown in Eq. 5.2.

$$FrameLossRate = \frac{FR_{enc} - Avg_{fps}}{FR_{enc}} \quad (5.2)$$

Frame losses in the video stream caused by packet loss can influence the gaming QoE by affecting the visual perception of players, which in consequence can also interrupt their interaction with the game due to delayed visual feedback, explained

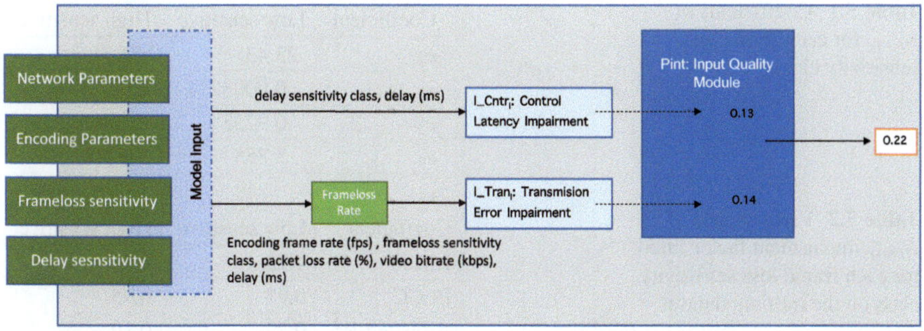

Fig. 5.1 Input quality module of the gaming QoE model, adapted from Fig. 3.1

by the impairment factor I_{trans_I}. The effect on the visual perception is measured based on the video discontinuity, whose effect is modeled as a part of the video transmission error impairment factor, I_{trans_V}. This impairment factor is described in more detail in Sect. 5.2.

In this section, the consequences of delayed visual feedback from the game are considered. Therefore, the input quality is measured and modeled as a part of the (I_{trans_I}). Additionally, the control latency impairment factor, I_{cntr_I}, is introduced that describes the impact of delay on input quality. The two impairment prediction models are trained based on the relevant conditions in the interactive dataset that was introduced in Sect. 3.4.5. The performance of the fitted models for each impairment is reported as "goodness of fit" and evaluated in terms of RMSE, PLCC, and R-squared measurements. It has to be noted that in order to show the goodness of fit, all performance metrics are reported on the R-scale and are not transformed back to MOS.

Figure 5.1 illustrates the gaming input quality module adapted from Fig. 3.1 that is briefly presented in Chap. 3. The figure also shows the input parameters of each impairment model. The two impairment models are integrated into a single diagnostic output, $O.22$, to predict the input quality, which is discussed at the end of the chapter. In the following sections, the developed model for each impairment is described.

5.1.1 *Effect of Transmission Error Impairment on Input Quality*

In order to model the impairment factor, I_{tran_I}, all input quality ratings (cf. Sect. 3.4.5) from the interactive dataset for conditions with varying framerates and packet loss were used. For the impairment factor I_{tran_I}, the model is developed based on the two parameters of $FrameLossRate$ and encoding framerate, FR_{enc}. Equation 5.3 is used for the prediction of I_{tran_I}. The values of the coefficients for

Table 5.1 Coefficients of I_{tran_I} for each frame loss sensitivity class

Coefficient	Low sensitive	High sensitive
e_1	23.43	54.71
e_2	0.0008574	0.02589
e_3	−0.9253	−2.485
e_4	5.855	9.306

Table 5.2 Performance of I_{tran_I} impairment factor fitted for each frame loss sensitivity class on the training dataset

Coefficient	Low sensitive	High sensitive
RMSE	4.23	4.98
PLCC	0.91	0.95
Adjusted-R^2	0.79	0.89

each frame loss sensitivity class are presented in Table 5.1. It has to be noted that while the encoding framerate is not a transmission error, due to the similarity of the effect on the input quality, it is decided to integrate it in the transmission impairment modeling. The lowest framerate level that is tested on the interactive dataset was 10 fps. For training the model, approximately 60% of the games in the interactive dataset, including six video games from different classes of delay sensitivity, are used.

$$I_{tran_I} = e_1 + e_2 \cdot FR_{enc}^2 - e_3 \cdot FR_{enc} + e_4 \cdot log(FrameLossRate + 1) \quad (5.3)$$

The performance of the impairment factor I_{tran_I} on the training dataset is reported in Table 5.2 by means of RMSE on R-scale level and PLCC as well as adjusted R-square. It can be observed from the table that the prediction model correlates well with subjective ratings on the training set for both high and low sensitive classes. However, the model fits better for the high sensitive class, which might be due to higher variability of ratings in this sensitivity class.

5.1.2 Effect of Control Latency Impairment on Input Quality

For the calculation of the control latency impairment, I_{cntr_I}, the parameter network delay is used. The coefficients are trained based on the training data collected in the interactive dataset and sensitivity of games towards delay. In the development of the interactive dataset, the network delay was introduced artificially by a Linux traffic control tool, NetEm.[1] As discussed in the previous section, other types of delay that might occur between client and server are not considered for the model with the assumption that those additional delays are comparatively low and also constant in a laboratory setup. Equation 5.4 is derived to predict the impairment I_{cntr_I}. The coefficients obtained for each delay sensitivity class are summarized in Table 5.3.

[1] http://manpages.ubuntu.com/manpages/trusty/man8/tc-netem.8.html.

Table 5.3 Coefficients of I_{cntr_I} for each delay sensitivity class

Coefficient	Low sensitive	High sensitive
f_1	47.97	90
f_2	2.097	1.191
f_3	0.01073	0.009775
f_4	−4.567	−18.73

Table 5.4 Performance of I_{cntr_I} impairment factor fitted for each delay sensitivity class on the training dataset

Coefficient	Low sensitive	High sensitive
RMSE	2.98	8.99
PLCC	0.98	0.96
Adjusted-R^2	0.96	0.90

$$I_{cntr_I} = \frac{f_1}{1 + exp(f_2 - f_3 \cdot Delay)} + f_4 \tag{5.4}$$

The performance of the impairment factor I_{cntr_I} on the training dataset is reported in Table 5.4 by means of RMSE on R-scale and PLCC as well as adjusted R-square. While the performance is good for both classes, it can be observed that the RMSE is noticeably lower for the low delay sensitivity class. This might be due to a smaller spread of the input quality data among the corresponding games as the effect of delay on the interaction quality is minor for the low complexity class within the range of delay values (between 0 ms to 400 ms) tested in the interactive dataset.

5.2 Effect of Video Transmission Error Impairment on Video Quality

In order to model the impairment factor, I_{Trans_V}, all discontinuity ratings from the interactive dataset for conditions with varying framerates and packet loss are used. It must be noted that the video discontinuity does only represent the level of jerkiness that a participant perceives. It does not consider its impact on the input quality. As an example of such a difference between these two quality aspects in the interactive dataset, the input quality ratings (mean value of the GIPS items) and the video discontinuity ratings for one game are compared. The selected game, Tekken 7, is a very fast-paced game, controlled only by keyboard inputs. Therefore, due to a fast-paced scene, the freezing effect due to packet loss and low encoding framerate is strongly visible to the participant which leads to low discontinuity ratings. However, this does not strongly affect the interaction of participants with the game, which might be due to the very frequent input events performed by a user resulting in a reduced perception of the temporal feedback from the game. Figure 5.2 illustrates the barplots comparing the discontinuity and input quality ratings for the game Tekken in the interactive dataset for different levels of packet loss and framerate

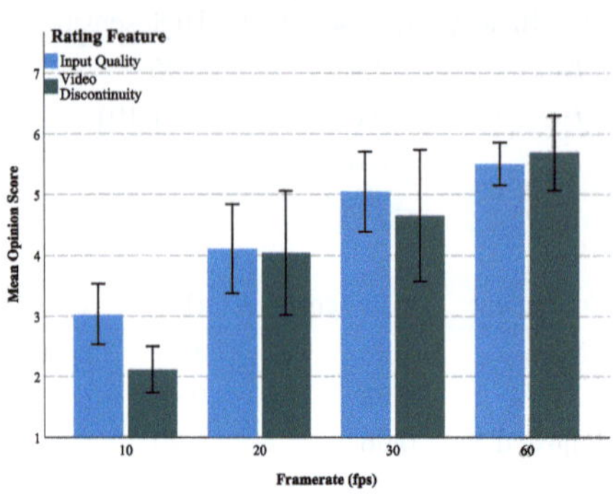

(a)

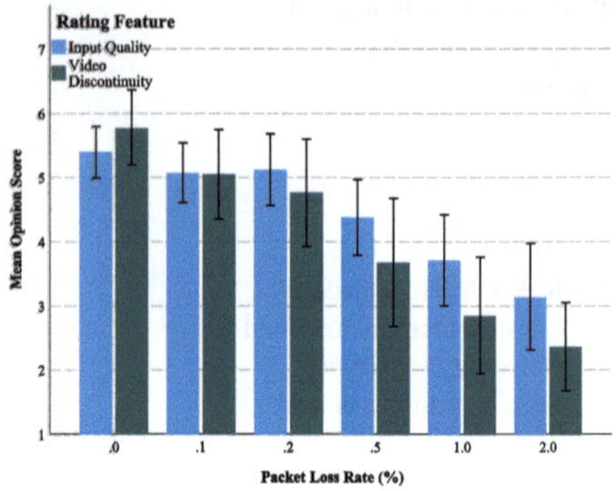

(b)

Fig. 5.2 Barplot of input quality and video discontinuity for the game Tekken in interactive dataset (c.f. Chap. 3). (**a**) Framerate variation for the game Tekken. (**b**) Packet loss variation for the game Tekken

Table 5.5 Coefficients of I_{Tran_V} for each frame loss sensitivity class

Coefficient	Low sensitive	High sensitive
g_1	29.13	47.03
g_2	0.01344	0.01747
g_3	−1.283	−1.823
g_4	6.724	10.7

when all other influencing factors remained at the highest values, i.e., $bitrate = 50$ Mbps and $resolution = 1080$p.

It can be observed that video discontinuity is differently impacted by packet loss and low encoding framerates than the input quality. Thus, depending on the scenario sensitivity, it appears to be reasonable to distinguish between the effect packet losses have on the input quality and of their impact on the video discontinuity for modeling gaming QoE.

Similar to I_{Tran_I}, the impairment factor I_{Tran_V} also uses the two parameters of $FrameLossRate$ and FR_{enc}. Equation 5.5 was derived for the prediction of I_{Tran_V} according to the frameloss sensitivity class. The coefficients with respective values for each frame loss sensitivity class are presented in Table 5.5.

$$I_{Tran_V} = g_1 + g_2 \cdot FR_{enc}^2 - g_3 \cdot FR_{enc} + g_4 \cdot log(FrameLossRate + 1) \quad (5.5)$$

The goodness of fit for the impairment factor I_{Tran_V} on the training dataset is reported in Table 5.6 by means of RMSE on R-scale and PLCC as well as adjusted R-squared. The result reveals a good fit for both low and high complexity classes where the performance is slightly higher for low complex class.

Table 5.6 Performance of I_{Tran_V} impairment factor fitted for each video complexity class on the training dataset

Coefficient	Low complex	High complex
RMSE	6.48	9.97
PLCC	0.85	0.81
Adjusted-R^2	0.71	0.64

Table 5.7 Weighting factors of the core gaming QoE model

Coefficient	*a*	*b*	*c*	*d*
Value	0.68	0.15	0.23	0.66

5.3 Core Gaming QoE Model

The core model predicting gaming QoE based on the impairment factor approach was defined in Chap. 3. Its structure can be described by Eq. 5.6.

$$Q_{Gaming} = Qo_{Gaming} - a \cdot I_{cod_V} - b \cdot I_{trans_V} - c \cdot I_{trans_I} - d \cdot I_{cntr_I} \tag{5.6}$$

The core model is trained based on the interactive dataset which is used to derive the coefficients that are presented in Table 5.7. As it can be seen from the coefficients, the weight for the I_{tran_V} is much lower than the remaining weights. A possible explanation is that most of the variance of I_{tran_V} on the overall gaming QoE is already covered by I_{tran_I}.

Based on the collected data from the interactive dataset, MOS_{max}, the MOS value corresponding to Qo_{Gaming}, is 4.64, whereas the lowest possible MOS value, MOS_{min}, is 1.3. Thus, the MOS_{QoE} can be derived using $MOSfromR$ transformation with the input of measured Q_{Gaming}.

$$MOS_{QoE} = MOSfromR(Q_{Gaming}) \tag{5.7}$$

The transformation of the R-scale to the MOS scale can be calculated according to the transformation provided in ITU-T Rec. P.1201.2 [113] as described in Algorithm 1.

In addition to the core gaming QoE prediction, the proposed framework provides diagnostic information predicting the input quality and video quality. In order to predict the diagnostic video quality output, O_{21}, and considering multiple video coding impairment models (cf. Chap. 4), the following approach is considered. First, the ratings of two video quality dimensions, fragmentation, and unclearness, are used to predict the impairment I_{cod_V} with a linear model for all conditions for which the video discontinuity is not degraded. Next, the predicted I_{cod_V} and video discontinuity ratings in the interactive dataset are used to predict the O_{22}. Training the model following this approach allows to avoid being biased to a certain type of model proposed for I_{cod_V} in Chap. 4. However, it has to be noted that using only fragmentation and unclearness to predict I_{cod_V} might introduce small errors due to neglecting the low encoding framerate affecting coding impairment. However, this

Algorithm 1 MOSfromR

procedure MOSFROMR(Q_{Gaming})
  $MOS_{min} \leftarrow 1.3$
  $MOS_{max} \leftarrow 4.64$
  if $Q_{Gaming} > 0$ *and* $Q_{Gaming} < 100$ **then**
    $MOS_{QoE} \leftarrow (1 + (MOS_{max} - MOS_{min})/100 \cdot Q_{Gaming} + Q_{Gaming} \cdot (Q_{Gaming} - 60) \cdot (100 - Q_{Gaming}) \times 7.0e - 6)$
  else
    if $Q_{Gaming} \geq 100$ **then**
      $MOS_{QoE} \leftarrow MOS_{max}$
    end if
    if $Q_{Gaming} \leq 0$ **then**
      $MOS_{QoE} \leftarrow MOS_{min}$
    end if
  end if
end procedure

effect is counterbalanced by I_{trans_V} for prediction of O_{21}.

$$O_{21} = Qo_{Video} - 0.93 \cdot I_{cod_V} - 0.43 \cdot I_{trans_V} \tag{5.8}$$

The diagnostic input quality output, O_{22}, can be predicted based on the estimation of two impairments of I_{trans_I} and I_{cntr_I} following Eq. 5.9. It has to be noted that O_{22} is predicted based on the prediction values of I_{trans_I} and I_{cntr_I} discussed in the previous sections, and input quality ratings as target value from the interactive dataset. The linear model is built after excluding irrelevant conditions to input quality, e.g., if only the video coding impairment is present. In contrast, for the core gaming QoE prediction model, all conditions are taken into account to predict the overall quality ratings. This explains the difference between the coefficients of the two equations.

$$O_{22} = Qo_{Input} - 0.83 \cdot I_{trans_I} - 0.94 \cdot I_{cntr_I} \tag{5.9}$$

5.4 ITU-T Rec. G.1072

In the following section, a short overview of the standardized opinion model predicting gaming QoE, ITU-T Rec. G.1072 [51], is given, to which the author of this book contributed to between 2016 to 2020.

ITU-T Rec. G.1072 is a planning model predicting cloud gaming quality, which is developed in a very similar approach that is described in this book. However, there are a few differences that will be described in this section. To begin with, Fig. 5.3 illustrates the model structure of ITU-T Rec. G.1072.

Similar to the approach in the book, the model structure is composed of two main modules: input quality (IPQ), and video quality (VQ). The VQ module is predicted

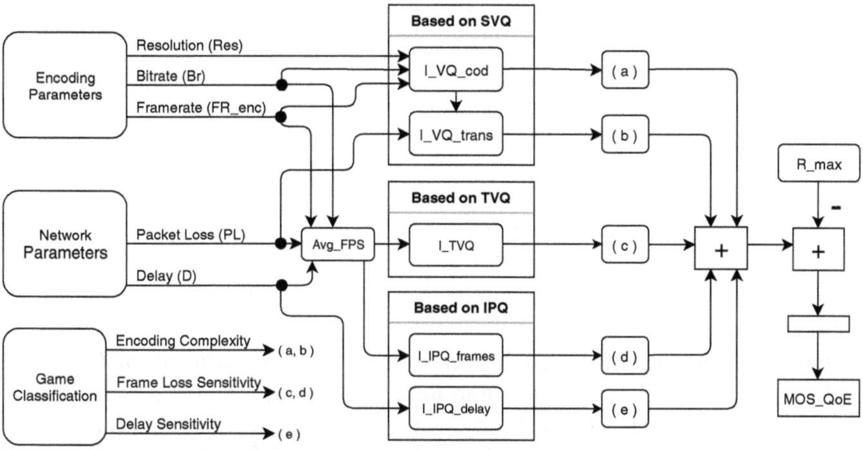

Fig. 5.3 Model structure of ITU-T Rec. G.1072 taken from [51], figure G.1072(20)_F01

based on the I_{VQ-cod}, $I_{VQ-trans}$, and I_{TVQ}. The first two build the spatial video quality, and the latter represents temporal video quality. I_{TVQ} represents the I_{tran_V} in the book that is measured by Eq. 5.5, and I_{VQ-cod} represents the I_{cod} in the book which is described in Sect. 4.1.1. However, one of the main difference between the model in the present book and G.1072 is the $I_{VQ-trans}$. $I_{VQ-trans}$ estimates the spatial video quality impairment by video transmission error if the transmission error is concealed by a PLC mechanism that introduces slicing artifacts. The packet loss concealment method that introduces the slicing artifacts is not practically implemented in any of the well-known cloud gaming services, as discussed earlier in this chapter. Thus, it was not considered in this book. For the calculation of the impairment factor, $I_{VQ-Trans}$, the parameters packet loss (PL) and results from I_{VQ-Cod} are used, following multiple equations presented below which are inherited from the ITU-T Rec. G.1071 [112] and ITU-T Rec. P.1201.2 [113]. The impairment $I_{VQ-Trans}$ is measured based on the $LossMagnitudeE$ that captures the packet loss degradation when slicing is applied as packet loss concealment. It has to be noted the burstiness of loss is not considered and set to zero in the ITU-T Rec. G.1072.

$$I_{VQ-Trans} = c_{1V} \cdot log(c_{2V} \cdot LossMagnitudeE + 1) \quad (5.10)$$

$$LossMagnitudeE = q_1 \cdot exp(q_2 \cdot LossMagnitudeNP) - q_1 \quad (5.11)$$

$$LossMagnitudeNP = \frac{(c_{21} - I_{codn}) \cdot PL}{c_{23} \cdot I_{codn} + PL} \quad (5.12)$$

Table 5.8 Weighting factors of the ITU-T Rec. G.1072 core model

Coefficient	*a*	*b*	*c*	*d*	*e*
Value	0.788	0.896	0.227	0.625	0.848

$$I_{codn} = \begin{cases} I_{VQ_{cod}} & \text{if } I_{VQ_{cod}} \leq 65 \\ 65 & \text{else} \end{cases} \tag{5.13}$$

where PL represents the percentage of lost transport stream video packets in the measurement window.

Finally, the input quality module is composed of two impairments of $I_{IPQ_{frames}}$ and $I_{IPQ_{delay}}$ that are corresponding to the I_{Tran_I} and I_{cntr_I} in this book, respectively. The core model of G.1072 follows a very similar structure but is taking into account one additional impairment factor. The core model predicting gaming QoE based on G.1072 is defined as in Eq. 5.14 and coefficients are presented in Table 5.8

$$\begin{aligned} Q_{Gaming} = Q_{Gaming} - a \cdot I_{VQ-cod} - b \cdot I_{VQ-Trans} - c \cdot I_{TVQ} \\ -d \cdot I_{IPQ_{frames}} - e \cdot I_{IPQ_{delay}} \end{aligned} \tag{5.14}$$

In the development of ITU-T Rec. G.1072 the interactive dataset, as well as CGVDS together with another dataset that covers the slicing effects are used. Therefore, while there are some minor differences in the model development and impairment names, the core idea of the modular model development is the same.

5.5 Summary

In this chapter, three prediction models are presented for three impairment factors in the gaming QoE model:

- I_{cntr_I} predicts the effect of control latency impairment on the input quality. The effect is modeled based on the network parameter of delay.
- I_{tran_I} predicts the effect of transmission error impairment on the interaction quality. This impairment is modeled based on the encoding framerate and $FrameLossRate$.
- I_{tran_V} predicts the effect of transmission error impairment on the video quality perception. This impairment similar to I_{tran_I} and is modeled based on the encoding framerate and $FrameLossRate$.

Packet loss in cloud gaming services typically results in frameloss, called freezing effect. These framelosses affect the visual perception as well as interaction of players. Thus, two impairment factors are presented to model the impact of packet loss on the input and video quality separately.

In addition, the development of the core gaming QoE model is described. Furthermore, the structure of the opinion model ITU-T Rec. G.1072 was presented that follows closely the structure of the core model. However, the ITU-T Rec. G.1072 uses one more impairment factor to model the effect of packet loss on (spatial) video quality when using a concealment approach which can result in slicing artifacts.

In the next chapter, the performance of the developed models for the prediction of gaming QoE is presented for passive and interactive datasets in terms of PLCC, SRCC and RMSE.

Chapter 6
Performance Evaluation

6.1 Evaluation of Video Coding Impairment

In this section, the proposed video coding impairment prediction models are evaluated based on three passive gaming datasets introduced in Sect. 3.4. In order to evaluate these models, the performance is measured based on the video quality MOS for the sake of simplicity in comparison with other existing video/image quality models. Thus in this chapter, these models are referred to as video quality prediction models. However, as discussed in Chap. 4, the result of these model can be transformed to the R-scale. For each type of model, detailed information about the performance of each model is given using scatter plots followed by a discussion on their performance individually. The section is organized by the type of model according to the video quality model classification presented in Sect. 2.6.

In order to get a deeper insight into the performance of each proposed model, they are compared to a total of six well-known existing image/video quality prediction models as described below.

Peak Signal to Noise Ratio (PSNR) is a widely used video quality metric and relies on the computation of the logarithmic difference between corresponding pixels in the original and impaired frames.

Structural Similarity Index Metric (SSIM) measures the structural similarity between two images and usually provides better video quality predictions compared to PSNR.

Video Multi-Method Assessment Fusion (VMAF) developed by Netflix, fuses three different metrics together to obtain a single score as the estimation of the video quality [74].

Blind/Referenceless Image Spatial Quality Evaluator (BRISQUE) uses locally normalized luminance coefficient and tries to quantify the possible loss of "naturalness" [95].

S. Zadtootaghaj, *Quality of Experience Modeling for Cloud Gaming Services*,
T-Labs Series in Telecommunication Services,
https://doi.org/10.1007/978-3-030-98249-2_6

Natural Image Quality Evaluator (NIQE) is based on a space domain NSS model, and is a learning-based quality estimation metric [94].

Perception based Image Quality Evaluator (PIQE) is an NR metric that uses cues from the human visual system [114].

In addition to the mentioned metrics, the quality predictions of ITU-T P.1203 and P.1204.3, which were described in Chap. 4, are also reported. For the computation of PSNR and SSIM, the VQMT tool is used which is available in [115]. For VMAF calculation, the Linux based implementation made available by the developers in [74] (version: $VMAF-VF0.2.4b-0.6.1$) is used. The built-in MATLAB function is used to measure PIQE, NIQE, and BRISQUE (version: R2019a).

6.1.1 *Planning Models*

In this section, the performance of four planning models is evaluated based on three different gaming video datasets. It has to be noted that the ITU-T Rec. P.1203 Mode 0 is considered as a planning model and evaluated in this section as it uses the same level of information as other proposed models in this section.

Table 6.1 shows the performance of the planning models in terms of PLCC, SRCC, and RMSE for three gaming video datasets. In order to report the video quality of G.1071 and G.1072, the video quality is measured only based on the coding impairment, I_{cod}, of these two models, and after transforming back the R-scale to a 5-point MOS scale. It has to be noted that GamingPara is trained based on the CGVDS. Thus, the performance of GamingPara is only reported on the validation part of the CGVDS but not on the whole dataset.

Based on the performance metrics reported in Table 6.1, it can be observed that GamingPara outperforms the other models in terms of correlation considering the

Table 6.1 Performance of planning models on three gaming video quality datasets of GVSET, KUGVD, and CGVDS

		Planning models			
Dataset	Metrics	G.1071	G.1072[a]	P.1203 Mode 0	GamingPara[a]
GVSET dataset	PLCC	0.68	0.77	0.30	0.83
	SRCC	0.65	0.81	0.35	0.82
	RMSE	1.10	1.09	1.19	0.69
KUGVD dataset	PLCC	0.73	0.78	0.19	0.87
	SRCC	0.70	0.72	0.26	0.87
	RMSE	0.99	0.68	1.24	1.08
CGVDS dataset	PLCC	0.60	0.73	0.66	0.78
	SRCC	0.56	0.69	0.69	0.75
	RMSE	0.68	0.48	0.88	0.39

[a] The performance of GamingPara and G.1072 are only reported based on the validation dataset of CGVDS (cf. Chap. 4)

three gaming video quality datasets. However, GamingPara results in high RMSE for the KUGVD and GVSET datasets, which will be discussed in the next section. Surprisingly, ITU-T Rec. P.1203 Mode 0 performed poorly for the KUGVD and GVSET datasets, while the performance is reasonable for the CGVDS. It needs to be noted that the CGVDS has a higher diversity of encoding parameters compared to the other two datasets which might play an important role in the improvement of the performance for P.1203 Mode 0.

As discussed in Chap. 4, ITU-T Rec. G.1072 uses the ITU-T Rec. G.1071 model structure to predict the coding impairment. However, ITU-T Rec. G.1072 is fitted again based on the training data of the CGVDS for each class of video complexity. This process considerably improves the performance of G.1072 for the CGVDS test datasets compared to G.1071.

6.1.1.1 Insights into the Performance of Planning Models

The proposed GamingPara is a planning model that is trained based on a partial dataset of CGVDS. In the development of CGVDS, the NVENC codec was used for encoding the videos using the llhq preset. The llhq preset does not use B-frames which leads to higher requirements of bitrate compared to the case that an encoder is allowed to use B-frames to reach a similar level of quality. Therefore, when GamingPara is tested on videos encoded with a preset that uses B-frames, GamingPara might underestimate the quality. Another interesting observation regarding the llhq preset is reported in [116]: the llhq preset uses multiple passes for encoding, which leads to a significant improvement of bit allocation for very low bitrate levels.

In the two gaming video datasets, KUGVD and GVSET, the software implementation of H.264 with the *"veryfast"* preset is used. Thus, the GamingPara generally underestimates the quality due to the selection of the *"veryfast"* present compared to the bitrate demanding llhq preset. However, in low bitrate ranges, the result shows that the model even overestimates the quality, which is expected as discussed earlier. Although the correlation analysis shows a promising result of GamingPara for GVSET and KUGVD according to Table 6.1, the scatter plot presented in Fig. 6.1 shows how GamingPara under/overestimate the quality for KUGVD and GVSET. Similarly, the video coding impairment of ITU-T Rec G.1072 is trained based on a part of CGVDS as well as another passive video quality dataset that are both using hardware implementation of H.264/MPEG AVC (using NVENC and llhq preset). Thus, similar to the GamingPara result, as it can be seen from Table 6.1, the model correlates well on GVSET and KUGVD but result in relatively high RMSE on these two datasets.

This is one of the main drawbacks of the planning models: they depend strongly on simple parameters of encoding, which by changing the codec or encoding settings lead to the requirement of retraining the models to yield valid results.

The scatter plot of GamingPara on both training and validation parts of CGVDS, illustrated in Fig. 6.2, shows an auspicious result considering the fact that this model only uses very basic video encoding parameters. However, it has to be noted

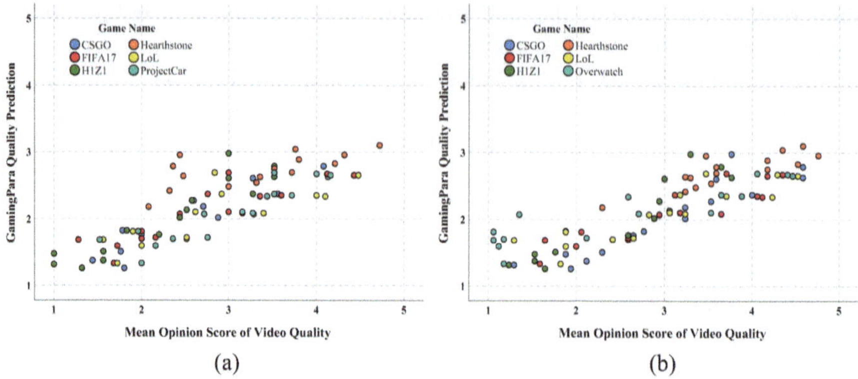

Fig. 6.1 Scatter plots of predicted GamingPara and assessed video quality for the two test datasets of GVSET and KUGVD. (**a**) Predicted GamingPara vs. MOS on GVSET dataset. (**b**) Predicted GamingPara vs. MOS on KUGVD dataset

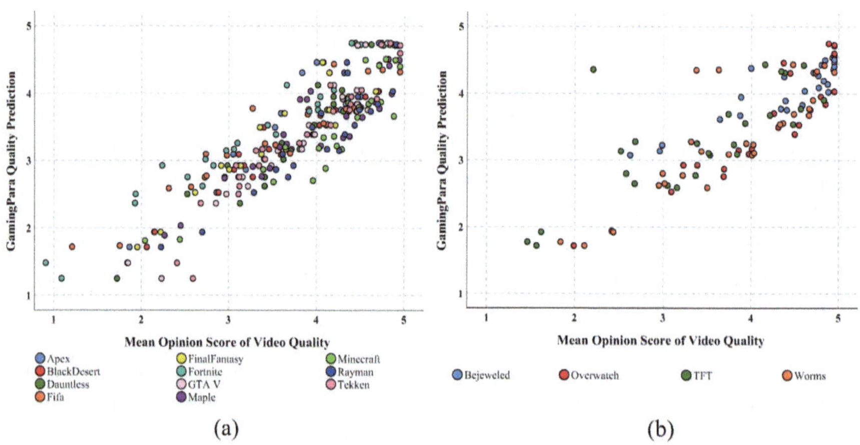

Fig. 6.2 Scatter plots of predicted GamingPara and assessed video quality on CGVDS dataset. (**a**) Predicted GamingPara vs. MOS on training CGVDS dataset. (**b**) Predicted GamingPara vs. MOS on validation CGVDS dataset

that GamingPara relies on the gaming video complexity classification. Thus, if the information about the classification is not available or the scene is not representative of the complexity class, it is expected that the model performs worse.

6.1.2 Bitstream-Based Models

In this section, the performance of the proposed BQGV model is compared to the ITU-T Recommendation models, P.1203 and P.1204.3. The performance of the

Table 6.2 Performance of bitstream-based models on CGVDS

Bitstream based models					
Metrics	P.1203 m1	P.1203 m3	P.1204.3	FHD P.1204.3	BQGV[a]
PLCC	0.58	0.88	0.84	0.84	0.90
SRCC	0.62	0.88	0.85	0.84	0.89
RMSE	0.69	0.48	0.58	0.40	0.33

[a] BQGV is trained and evaluated following a cross-validation with a split into five bins which in each bin three games are randomly selected

bitstream-based models is reported in terms of PLCC, SRCC and RMSE in Table 6.2 on the CGVDS. The result reveals a high performance of the BQGV, ITU-T Rec. P.1203 Mode3, and P.1204.3 models. As discussed in Sect. 4.2, BQGV is trained based on the CGVDS. To train and test the model fairly, the CGVDS is split into five bins which in each bin three games are randomly selected. The model was trained five times, each time one bin was hold-out and the model was trained based on the remaining videos. Therefore, the performance evaluation of BQGV presented in Table 6.2 is based on multiple iterations for the CGVDS dataset.

On the CGVDS, the BQGV performs similar to the ITU-T models P.1203 Mode 3 and P.1204.3. However, it must be noted that BQGV might be biased to the encoding setting of CGVDS, whereas the two other standardized models are trained on video quality datasets that did not use the encoding setting following CGVDS dataset (e.g., llhq preset). The high performance of the two ITU-T models on CGVDS dataset is because of using packet payload information that includes detailed information such as assigned quantization parameters. In contrast, BQGV only uses the packet header information.

The proposed BQGV performs poorly compared to the two other ITU-T models on the GVSET, as indicated by a PLCC of 0.62 compared to 0.87 and 0.89 for P.1203 and P.1204.3 FHD, respectively. This poor performance is due to the training process of BQGV that has been done on CGVDS, which uses llhq preset. Since there is an infinite GoP for the llhq preset, the bitstream pattern is different in CGVDS compared to KUGVD and GVSET which leads to low performance of the BQGV model on datasets that are created using closed GoP. Thus, the model does not perform well on KUGVD and GVSET.

Finally, the ITU-T Rec. P.1203 Mode 1 turned out to perform poorly with all three gaming datasets. ITU-T Rec. P.1203 Mode 1 is comparable to BQGV in terms of level of bitstream-based information that it has access to. The reason behind the low performance of this model might be the difference of encoding and type of content that the model is train on.

To conclude, bitstream-based models that only use information from the packet header are susceptible to changing the encoding setting that they are being trained on. However, this type of bitstream-based model is much simpler than payload-based models. Thus, they are better suitable for monitoring purposes.

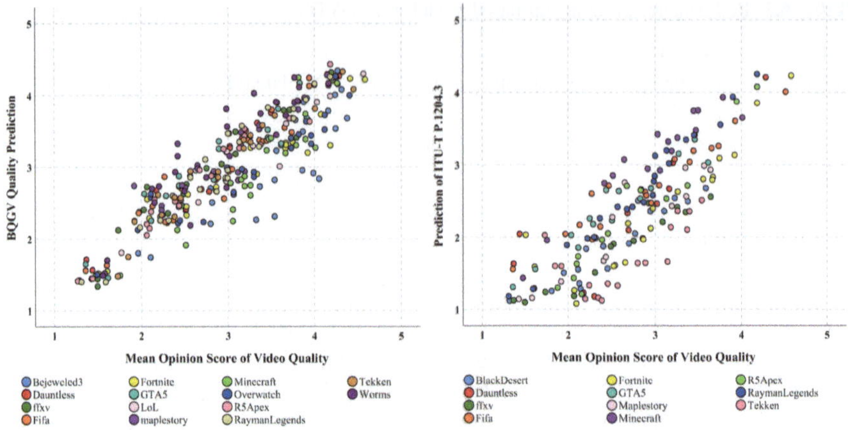

(a) (b)

Fig. 6.3 Scatter plots of predicted BQGV and ITU-T Rec. P.1204.3 compared to Mean Opinion Scores on the CGVDS test dataset. (**a**) Predicted BQGV vs. Mean Opinion Scores on CGVDS dataset. (**b**) Predicted P.1204.3 vs. Mean Opinion Scores on CGVDS dataset

6.1.2.1 Insights into the Performance of Bitstream-Based Models

Figure 6.3a illustrates the scatter plot of the BQGV predictions compared to the assessed MOS of video quality on the CGVDS. The BQGV is trained in five iterations based on the 5-s interval duration that is extracted randomly from the CGVDS dataset. Thus, the scatter plot on the CGVDS gives a big picture of how the model works for unknown gaming videos with encoding similar to the CGVDS compression setting. It can be observed that the BQGV performs poorly for two low complexity video games, Bejeweled3 and Worms, which might be due to a limited number of low complexity video games in the training set. The performance result on the two other gaming video datasets was poorly, PLCC of 0.62 for GVSET and 0.67 on KUGVD, which is due to differences in the encoding setting CGVDS compared to KUGVD and GVSET.

Figure 6.3b illustrates the scatter plot of P.1204.3 on CGVDS before FHD mapping. A small shift for prediction values can be observed, which is due to the usage of different screen sizes in training for the model compared to the CGVDS. Thus an FHD mapping is applied to fix this shift. It can be observed in Table 6.2 that the RMSE improved by 0.18 points after the FHD mapping.

The ITU-T P.1203 Mode 3 performs very well on the CGVDS, while it performs poorly when the resolution is 480p or framerate is set at 20fps. Figure 6.4 shows the performance of the ITU-T P.1203 Mode 3 once for only video sequences encoded at a framerate of 20 fps and once for those encoded only at a resolution of 480p. This might be due to the used training dataset which may have a low number of video sequences encoded at low resolution and low framerate.

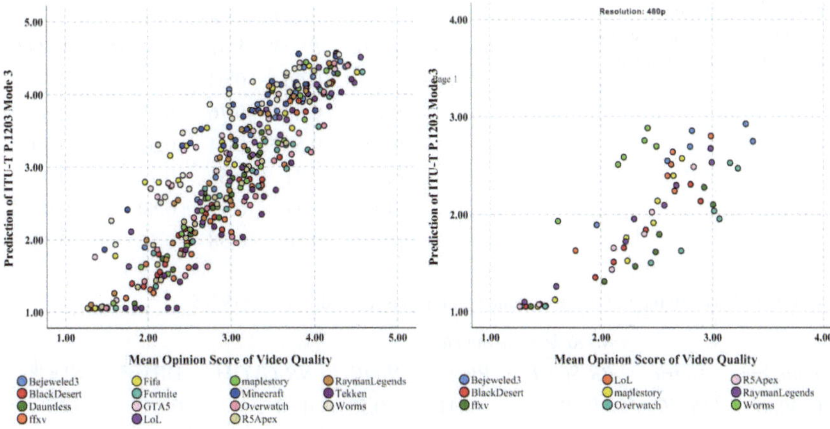

(a) (b)

Fig. 6.4 Scatter plots of predicted ITU-T Rec. P.1203 Mode 3 and assessed video quality on the CGVDS. (**a**) Predicted P.1203 vs. MOS on CGVDS at framerate of 20 fps. (**b**) Predicted P.1203 vs. MOS on CGVDS at resolution of 1080p

6.1.3 Signal-Based Models

In this section, the performance of the proposed signal-based models is evaluated and compared to the SoA video quality models. Since most of the SoA signal-based models and the proposed NR-GVQM do not predict the quality on a 5-point MOS scale, the prediction of those models are mapped to MOS values using a monotonous mapping function, a linear function or the more sophisticated monotonous part of a third order polynomial or a logistic function [117]. ITU-T Rec. P.1401 recommends such a mapping function to "compensates for offsets, different biases, and other shifts between the scores, without changing the rank order" [117]. The function is typically applied to the predicted scores before measurement of RMSE. In this book, for all metrics that predict the quality on a scale different than 5-point MOS scale, this mapping is applied. The RMSE is reported based on the lowest achieved RMSE among linear, third order polynomial and logistic mapping functions.

Tables 6.3 and 6.4 show the performance of FR and NR models in terms of PLCC, SRCC and RMSE for the CGVDS. Most of the SoA models are image-based quality metrics, and CGVDS is developed using different framerate levels. Considering the possible video discontinuity effect reflected on the subjective ratings due to variable framerate, the performance of metrics is evaluated once only at 60 fps for a fair comparison and once on all framerate levels. Surprisingly, for SSIM, the performance improves if all framerates are considered. This might be due to the fact that considering all framerate levels allows having more available data for the measurement of correlation, which in average results in higher correlation.

Table 6.3 Performance of Full-Reference signal-based models on CGVDS dataset

		Full Reference metrics			
Framerate	*Metric*	*PSNR*	*SSIM*	*MS-SSIM*	*VMAF*
20, 30, 60	PLCC	0.66	0.64	0.71	0.87
	SRCC	0.67	0.76	0.79	0.87
	RMSE	0.58	0.54	0.51	0.39
60	PLCC	0.71	0.61	0.73	0.91
	SRCC	0.71	0.72	0.80	0.90
	RMSE	0.56	0.58	0.51	0.33

Table 6.4 Performance of No-Reference signal-based models on CGVDS

		No Reference metrics					
Framerate	*Metric*	*BRISQUE*	*PIQE*	*NIQE*	*NR-GVQM*	DEMI[a]	NDG[b]
20, 30, 60	PLCC	−0.48	−0.41	−0.53	0.82	0.87	0.88
	SRCC	−0.46	−0.41	−0.55	0.81	0.87	0.88
	RMSE	0.71	0.74	0.66	0.47	0.49	0.53
60	PLCC	−0.45	−0.35	−0.51	0.85	0.86	0.89
	SRCC	−0.44	−0.36	−0.54	0.84	0.87	0.90
	RMSE	0.73	0.78	0.67	0.43	0.47	0.50

[a] Performance of DEMI is reported based on validation CGVDS
[b] NDG stands for NDNetGaming

In addition, the lowest framerate in the dataset is set at 20fps, which may not impact the video discontinuity significantly enough.

Based on the Table 6.4, it can be observed that the three NR metrics, PIQE, NIQE, and BRISQUE, perform poorly on the CGVDS, which is in line with the findings of [31]. It can be observed that NDNetGaming outperforms all NR metrics. It has to be noted that the performance of DEMI is reported only based on the validation part of CGVDS (cf. Chap. 4). In addition, for the prediction of NDNetGaming and DEMI in Table 6.4, the temporal pooling function is not applied, and frame-level predictions are pooled by simply averaging over frames prediction of each video sequence. The temporal pooling is not applied to have a fair comparison to other SoA models since SoA models do not have a temporal pooling function but rather a simple average pooling over frames of a video sequence (except VMAF). For NDNetGaming the frame-level quality is simply measured based on the output of the final trained CNN for each frame before applying the temporal pooling. To measure the frame-level quality for DEMI, the output of its second phase, i.e. the predicted unclearness and fragmentation after fine-tuning, is used. Based on the predicted unclearness and fragmentation and according to the linear regression fit presented in Sect. 3.4.4 (Eq. 3.5), the image quality is predicted. In the following section, the performance of NDNetGaming and DEMI using temporal pooling methods is discussed.

For two other gaming video datasets (KUGVD and GVSET), both NDNetGaming and DEMI models result in a very high correlation above 0.90 (for both PLCC and SRCC) based on the average pooling [90, 91]. The reader can refer to Sect. 2.7.3, Table 2.5, to find the performance of other signal-based models on the GVSET dataset.

6.1.3.1 Insights into the Performance of Signal-Based Models

No-Reference Gaming Video Quality Metric (NR-GVQM)

In order to get a deeper insight into the performance of NR-GVQM, the scatter plot of the predicted VMAF values (output of proposed NR-GVQM) and actual VMAF scores on the validation part of the GVSET dataset are presented in Fig. 6.5a, which achieves a PLCC of 0.97 and RMSE of 4.73. The RMSE is measured in the VMAF range of values vary from 20 to 95 on the GVSET. In addition, Fig. 6.5b illustrates the scatter plot of NR-GVQM predictions and assessed video quality MOS values, which follows a very similar trend to VMAF. It is important to note here that this analysis is performed on an unknown set of videos (in GVSET) to the trained model, and hence the results can be generalized for other gaming videos not considered in this dataset. A similar result is achieved for the KUGVD dataset indicated by a PLCC of 0.89 using MOS values. This was expected since the KUGVD and GVSET datasets are similar in content and encoding settings. However, NR-GVQM does not perform similarly well on the CGVDS. The NR-GVQM is trained based on the GVSET dataset that is created based on 30 fps encoding framerate, while the CGVDS is created using three encoding framerates of 20, 30, and 60 fps. Thus, the NR-GVQM over-predicts the videos encoded with 20 fps since it does not take into account the video discontinuity effect. Figure 6.6 illustrates the scatter plot of NR-GVQM prediction and MOS prediction for CGVDS, once at a framerate of 20 fps and once for the full dataset. Therefore, one of the main drawbacks of NR-GVQM is that it is only trained on a framerate of 30 fps, which require retraining for a lower framerate. It has to be noted that since NR-GVQM uses a temporal video feature, TI, as one of the input features, retraining the model could significantly improve the result for datasets created with variable framerates. Another drawback of the NR-GVQM is that it is trained based on VMAF values. That means at its best performance, it is expected to perform as well as VMAF. However, it must be noted that the main idea behind the development of NR-GVQM is to develop a model to close the gap between NR metrics and FR metrics for gaming content at a time which only one gaming video quality dataset was publicly available. One of the main advantages of NR-GVQM is that it does not require to have access to a subjective dataset. Thus, it can be retrained for new content, encoding settings, or higher performing objective metrics only by recording, encoding, and measuring the objective metric.

NDNetGaming

The performance analysis of this section is two-fold: at first, the performance of NDNetGaming is analyzed only based on the average pooling of frame-level

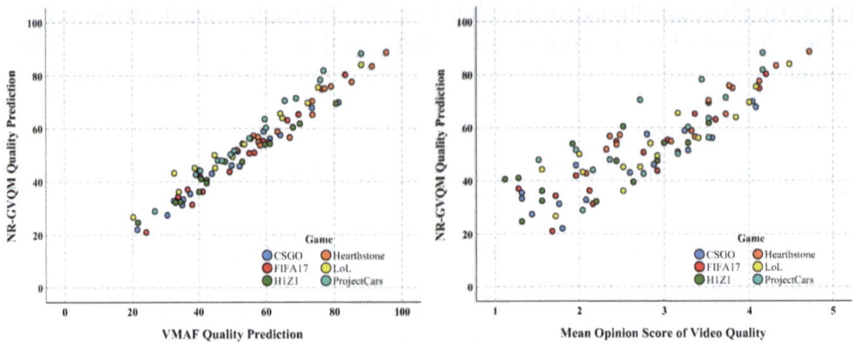

(a) (b)

Fig. 6.5 Scatter plots of predicted NR-GVQM vs. assessed video quality and VMAF for the GVSET test dataset. (**a**) Predicted NR-GVQM vs. VMAF on GVSET dataset. (**b**) Predicted NR-GVQM vs. MOS on GVSET dataset

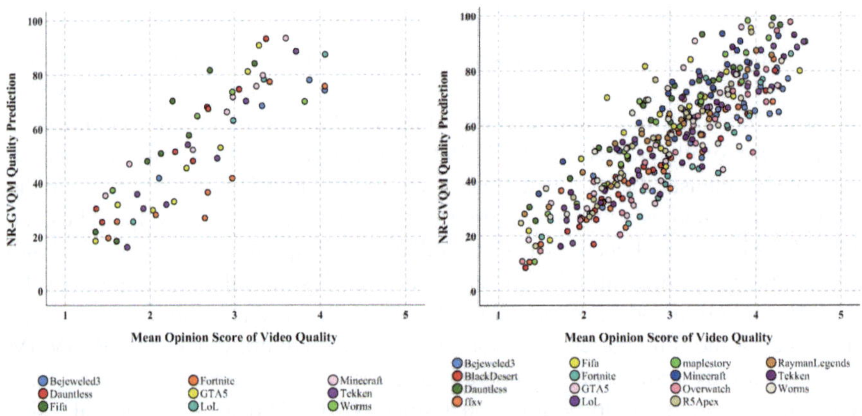

(a) (b)

Fig. 6.6 Scatter plots of predicted NR-GVQM and assessed video quality on the CGVDS, at framerate of 20 fps, as well as full dataset. (**a**) Predicted NR-GVQM vs. MOS on CGVDS at framerate of 20 fps. (**b**) Predicted NR-GVQM vs. MOS on CGVDS

predictions, and second, the performance of the model is investigated when the proposed temporal pooling function is applied.

Figure 6.7 shows the scatter plot of NDNetGaming predictions using the average pooling method and assessed video quality MOS ratings of GVSET and KUGVD. The model, using average pooling, reaches an RMSE of 0.347 and 0.464 for GVSET and KUGVD, respectively. Based on Fig. 6.7, a clear trend can be observed for complex games, e.g., Overwatch in KUGVD, where the NDNetGaming model underestimates the quality. However, for low complex video games, e.g., Hearthstone in GVSET, it overestimates the quality. This might be

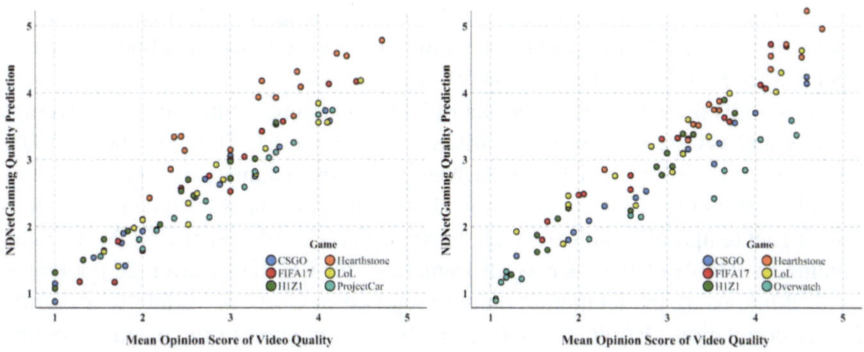

(a) (b)

Fig. 6.7 Scatter plots of predicted NDNetGaming using average pooling and assessed video quality on two datasets of GVSET and KUGVD. (**a**) Predicted NDNetGaming vs. MOS on GVSET. (**b**) Predicted NDNetGaming vs. MOS on KUGVD

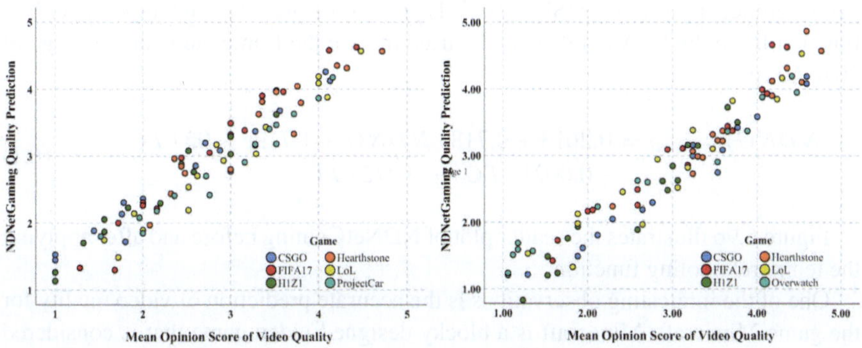

(a) (b)

Fig. 6.8 Scatter plots of predicted NDNetGaming using temporal pooling and assessed video quality on two datasets of GVSET and KUGVD. (**a**) Predicted NDNetGaming vs. MOS on GVSET. (**b**) Predicted NDNetGaming vs. MOS on KUGVD

due to a visual masking effect, i.e. in the presence of temporal or spatiotemporal distortions, the visibility of distortions to a human could strongly be reduced or completely removed [108]. Thus, even though the NDNetGaming might correctly detect the high spatial distortion in the high complexity video games, the effect of distortion is reduced due to visual masking by the human visual perception, reflected in the subjective ratings.

Next, the temporal pooling function was applied to the frame level prediction, and the result is presented in the form of scatter plots for KUGVD and GVSET in Fig. 6.8. It can be seen how the temporal pooling fixes the under/over-predictions that were presented previously in Fig. 6.7. After applying the temporal pooling, the result shows a PLCC of 0.961 (RMSE = 0.27) for GVSET and 0.968 (RMSE =

0.30) for KUGVD. One must consider that to investigate the effect of the temporal pooling method fairly, the GVSET is temporally pooled based on a linear fit that is obtained for KUGVD and vice versa (cf. Sect. 4.3.2).

Figure 6.9a presents the scatter plot of NDNetGaming predictions using average pooling and assessed video quality MOS values in the CGVDS. For the CGVDS, similarly to GVSET and KUGVD, NDNetGaming overestimates the quality for low complex games (e.g., Worms, Bejeweled3) and underestimates the quality for games with higher temporal complexity (e.g., Fortnite, GTA5, and Tekken). NDNetGaming obtains an RMSE of 0.53 across all framerates for CGVDS if average pooling is used. While using the temporal pooling proposed in Chap. 4, leading to a slightly higher correlation ($PLCC \approx SRCC = 0.89$) compared to the average pooling method, it increases the RMSE significantly to 0.77. The temporal pooling that was proposed in Chap. 4 is fitted only on the framerate of 30 fps, which fails to correctly map the videos with 20 fps and 60 fps framerates to MOS values of the CGVDS dataset. Therefore, the temporal pooling function is fitted again for CGVDS with the same function resulting in a slightly higher correlation ($PLCC \approx SRCC = 0.92$) but a significantly lower RMSE of 0.33. Equation 6.1 presents the temporal pooling function fit on the CGVDS that can be used in case the framerate is in the range of 20 to 60 fps.

$$\begin{aligned} NDNG_{Temporal} = &\ 0.2613 + 0.718 \cdot NDNG + 1.719e-05 \cdot TC^3 \\ &-0.0021 \cdot TC^2 + 0.072 \cdot TC \end{aligned} \tag{6.1}$$

Figure 6.9b illustrates the scatter plot of NDNetGaming before and after applying the temporal pooling function.

One of the interesting observations is the accurate prediction of video quality for the game Minecraft. Minecraft is a blocky designed video game that is considered as one of the most challenging scenes for quality prediction of SoA NR metrics. Several NR quality metrics struggle to accurately predict the quality of such a blocky video game, as the blocky design might be seen as an artifact. However, NDNetGaming also performs well for this game.

In addition to analyzing the scatter plots, it is interesting to visualize the intermediate feature layers of the NDNetGaming's CNN. Therefore, the local predictions of the model are visualized to get deeper insights into the function of intermediate feature layers and the operation of the CNN. To calculate the local predictions, a new network is built with the full-frame as an input (size 1080×1920) but with the same weights for the convolutional layers as the original model. Additionally, the global average layer between the convolutional and dense layer of the architecture is skipped. Usually, the dense layer expects a size of 1024 tensor as its input and calculates a single score as the output of the CNN. When skipping the global averaging layer, it instead gets 1024 33×60 matrices from the convolutional part of the network. The dense part of the model is applied element-wise to these matrices in order to get the local prediction matrix.

Figure 6.10 shows the local quality predictions for two frames in the GISET dataset. The local quality prediction is shown in a heat-map overlay layer that ranges

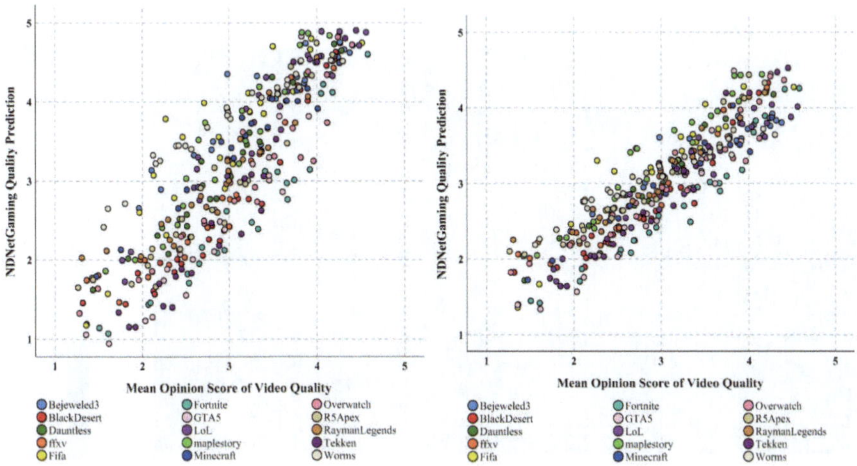

(a) (b)

Fig. 6.9 Scatter plots of predicted NDNetGaming and assessed video quality on the CGVDS. (**a**) Predicted average pooled NDNetGaming vs. MOS on CGVDS. (**b**) Predicted temporal pooled NDNetGaming vs. MOS on CGVDS

from red color, which represents high quality, to blue color, which represents low quality. In addition, another overlay layer is applied to show the 13-patch quality prediction design (cf. Chap. 4) for fast prediction of the quality. The quality labels in terms of MOS value, full-frame prediction, and 13-patch prediction are provided in the Fig. 6.10.

From the local quality predictions of many distorted images, it can be observed that the model has trouble distinguishing between the edges from blockiness and actual texture edges of the images. Therefore, when it deals with an image of high blockiness, the NDNetGaming model results in a high variation of local quality prediction as it can be seen in Fig. 6.10 for the game PlayerUnknown's Battlegrounds. However, on average, the image quality prediction does not deviate much from the subjective MOS score, considering the RMSE of 0.43 for images with blockiness degradation in the GISET dataset (1080p resolution and bitrate lower than 2 Mbps).

While there is a high variation of local quality predictions for blockiness artifacts, the model performs more consistently when dealing with blurriness artifacts, as it can be seen in Fig. 6.10 for the game League of Legends. As a general trend, it is observed that the developed model tends to predict the quality higher for regions with high texture complexity, which could also be in-line with human perception as image distortion can be masked in high texture regions of an image.

DEMI

DEMI gives both overall video quality as well as diagnostic predictions, which provide estimations of fragmentation and unclearness levels. Similar to the

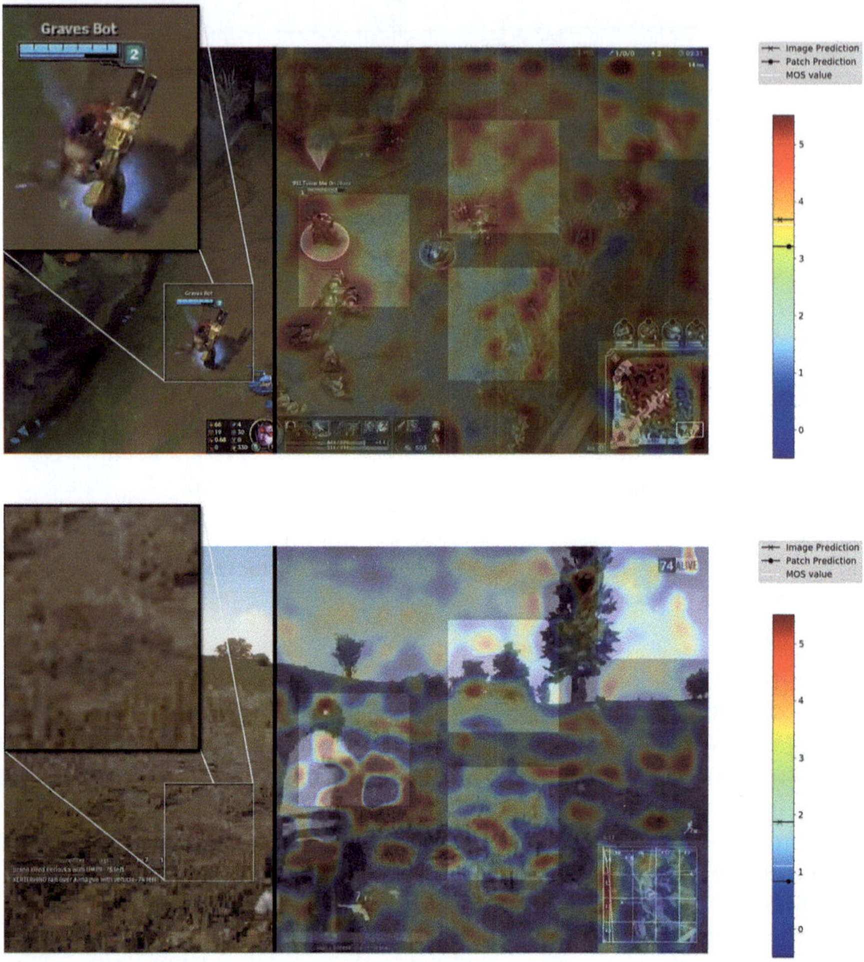

Fig. 6.10 Local quality predictions for one frame of League of Legends (top side) and PlayerUnknown's Battlegrounds (bottom side)

NDNetGaming approach, in order to fairly compare the performance of DEMI with other SoA models, the performance evaluation presented in Table 6.4 is reported only based on average pooling prediction. Since DEMI does not predict the image quality but rather the level of fragmentation and unclearness, a linear function (Eq. 3.5) that is fitted on GISET in Sect. 3.4.4.2 is used to map the unclearness and fragmentation prediction to image quality prediction. Using average pooling for prediction of DEMI result in a PLCC of 0.87 and RMSE of 0.49 for the CGVDS dataset as presented in Table 6.4.

Next, the temporal pooling is applied, which results in a significant improvement of prediction indicated by a PLCC, SRCC, and RMSE of 0.92, 0.93, 0.31,

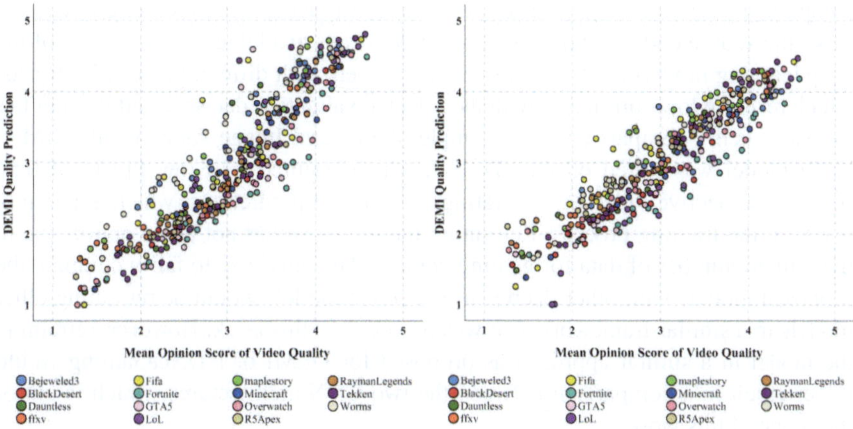

Fig. 6.11 Scatter plots of predicted DEMI and assessed video quality on the test dataset of CGVDS. (**a**) Predicted frame-level DEMI vs. MOS on CGVDS. (**b**) Predicted temporal pooled DEMI vs. MOS on CGVDS

respectively. Figure 6.11 presents the scatter plot of DEMI predictions using average pooling as well as using temporal pooling methods. It can be seen that the over/underestimations of the model are mapped closer to the subjective ratings for most cases. For the games Fifa and League of Legends, a few significant deviations from the subjective ratings can be seen, which could be due to the fact that the temporal pooling method is applied on every 20th frame.

Based on the NDNetGaming performance, it can be concluded that NDNetGaming performs slightly higher than DEMI for the CGVDS dataset. This could be due to the fact that NDNetGaming is trained only based on the gaming content and also uses the predictions of all frames of a video sequence compared to DEMI that predicts the quality based on every 20th frame.

6.1.3.2 Discussion

The performance of signal-based models was compared to a few well-known video/image quality metrics. Among the proposed NR models, NR-GVQM is a lightweight model that fuses the low-level image features to predict the gaming video quality. While a high performance of NR-GVQM has been seen for KUGVD and GVSET, the performance on the CGVDS is not similarly good. This might be due to the fact that this model does not take into account the temporal pooling aspect. However, the two proposed deep learning models, DEMI and NDNetGaming, perform similarly well if the temporal pooling is applied.

Since two of the proposed NR models are deep learning-based, it is interesting to compare their performance with SoA deep learning-based models. However, this comparison was not made due to the following practical and theoretical reasons. First, the source code of those models is not always available. Second, most of the deep learning models are trained based on datasets with different types of artifacts, which result in low correlation with the selected validation datasets, and it is not fair to make such a comparison. For example, the Neural Image Assessment (NIMA) [118] model was tested on the GVSET, which resulted in a low PLCC of 0.54 between subjective video quality ratings and model predictions, which apparently is due to the training process. Retraining the model would not be an option due to the limited number of data for gaming content. The only way to fairly compare the proposed models with other deep learning-based models would be retraining other models in a similar framework as it was proposed in this book. However, retraining the model in a similar approach as proposed for DEMI or NDNetGaming would rather result in a comparison between the two CNN architectures, which is out of the scope of this work.

In this section, the performance of the SoA signal-based models was compared to proposed gaming video quality metrics on the gaming video dataset. Such a comparison might not be fair since the SoA models are not designed/trained for gaming content. In addition, it is not clear how good the performance of SoA gaming quality metrics is compared to the proposed models. In the following, the performance of FR and NR models are evaluated on a non-gaming video quality dataset. In addition, the performance of the proposed gaming NR models will be compared to a few existing gaming video quality metrics in the literature.

6.1.4 Performance of Existing NR Gaming Video Quality Models

In this section, a short discussion on the performance of a few gaming video quality metrics in the literature is given. To the best of the author's knowledge, only three NR metrics are developed for/based on gaming content, as described in the following.

- *Lightweight NR Pixel Based Model for Gaming Content (nofu)* is a pixel-based video quality model for gaming content [62]. *Nofu* uses 12 different per frame features and a center crop approach for the fast computation of frame-level features.
- *NR-GVSQE* follows a similar approach as NR-GVQM, which is trained based on the multiple frame-level features as input features and VMAF values as target values.
- *NR-GVSQI* is an NR metric designed for gaming content trained based on a Neural Network [63]. It uses 15 low-level signal features, including the score of NR metrics such as BRISQUE and NIQE, for training the model.

Table 6.5 Performance of NR gaming quality metrics on GVSET and KUGVD datasets, in terms of PLCC and SRCC

	Metrics	Nofu	NR-GVSQE	NR-GVSQI	NR-GVQM	NDNetGaming	DEMI
GVSET	PLCC	0.91	–	0.87	0.89	0.93	0.91
	SRCC	0.91	–	0.86	0.87	0.93	0.91
KUGVD	PLCC	–	0.90	0.89	0.89	0.93	0.92
	SRCC	–	0.91	0.88	0.89	0.93	0.91

All three models are lightweight machine learning-based models that are trained based on GVSET or KUGVD. Interestingly, *nofu* takes into account some game-related features such as static area (e.g., head-up display in video games) as a feature to develop the model. At the time of *nofu*'s development, there was no public gaming video dataset other than GVSET to validate the model. Thus, a leave-one-out cross-validation was employed to evaluate the performance of the model. NR-GVSQI is trained two times, training based on GVSET then evaluating the model on KUGVD and vice versa. NR-GVSQE is only trained on GVSET and evaluated on KUGVD. Thus, the performance of *nofu* on KUGVD and *NR-GVSQE* on GVSET is not available.

Table 6.5 presents the performance of all NR gaming video quality models on KUGVD and GVSET based on the results reported in the reference papers. Based on the PLCC and SRCC values, it can be concluded that the two proposed deep learning-based models outperform the existing NR gaming quality models. *Nofu* performs higher than other lightweight gaming quality metrics on GVSET. However, it has to be noted that *nofu* uses leave-one-out cross-validation, whereas NR-GVQM and NR-GVSQI leave the whole part of GVSET that has subjective ratings out of the training. Also, it is not clear how these three gaming quality models work on a dataset with different encoding settings, e.g., on the CGVDS.

6.1.5 Performance on Non-Gaming Video Datasets

While the three proposed NR models are designed for gaming content, it is interesting to see how well they perform on non-gaming content. This allows drawing a conclusion about the potential performance of the proposed models on other types of content, e.g., a new gaming dataset. For this comparison, the NFLX-PD is selected that is described in Sect. 4.3.3. The selection of NFLX-PD is due to the similarity of encoding settings to typical gaming video streaming compression settings, e.g., encoding recommendation by Twitch.[1] Table 6.6 presents the result for the proposed NR model compared to the SoA image/video quality metrics.

[1] https://stream.twitch.tv/encoding/.

Table 6.6 Performance of FR and NR quality metrics on NFLX-PD dataset, in terms of PLCC, SRCC, and RMSE

	Full-reference models			No-reference models					
Metrics	PSNR	SSIM	VMAF	BRISQUE	NIQE	PIQE	GVQM[a]	DEMI	NDG[a]
PLCC	0.64	0.69	0.93	−0.77	−0.83	−0.78	0.82	0.89	0.89
SRCC	0.66	0.76	0.91	−0.76	−0.81	−0.80	0.80	0.89	0.89
RMSE	0.86	0.76	0.42	0.67	0.59	0.59	0.61	0.48	0.49

[a] GVQM stands for NR-GVQM and NDG stands for NDNetGaming

The results revealed a high performance of DEMI and NDNetGaming with MOS values on the NFLX-PD, while VMAF outperforms all metrics. However, it must be noted that VMAF is trained based on a similar dataset to NFLX-PD, and the result could be biased for VMAF on this dataset. The performance of DEMI and NDNetGaming summarized in Table 6.6 is measured based on the average pooling method. If the temporal pooling is applied, DEMI performs with $PLCC = 0.91$, $SRCC = 0.90$, and $RMSE = 0.43$ whereas NDNetGaming does perform slightly worse with $PLCC \approx SRCC = 0.87$ and $RMSE = 0.52$. This might be due to the fact that DEMI is trained based on both gaming and non-gaming datasets, while NDNetGaming is biased to gaming content. NR-GVQM performs quite well, considering the fact that it is trained based on VMAF.

6.2 Performance of Gaming QoE Model

In the previous section, the performance of the proposed coding impairment prediction models was presented. In this section, the performance of the remaining components of the gaming QoE model is evaluated, and, finally, the gaming QoE model is evaluated based on the interactive dataset.

While multiple gaming video quality datasets are developed for evaluation of video coding impairment, only one interactive dataset is created that follows the standardized quality assessment methodologies presented in ITU-T Rec. P.809. Therefore, the interactive dataset is divided into two parts of training and validation. The validation part of the model consists of ratings from four video games collected from four subjective studies. In the interactive dataset, for one game, CSGO, more video encoding parameters, e.g., more resolution levels and bitrates, are tested in the development of the interactive dataset.

6.2.1 *Input Quality and Video Discontinuity*

The input quality can be predicted based on the control latency impairment (I_{cntr_I}) and video transmission error impairment (I_{trans_I}). Since these two impairment

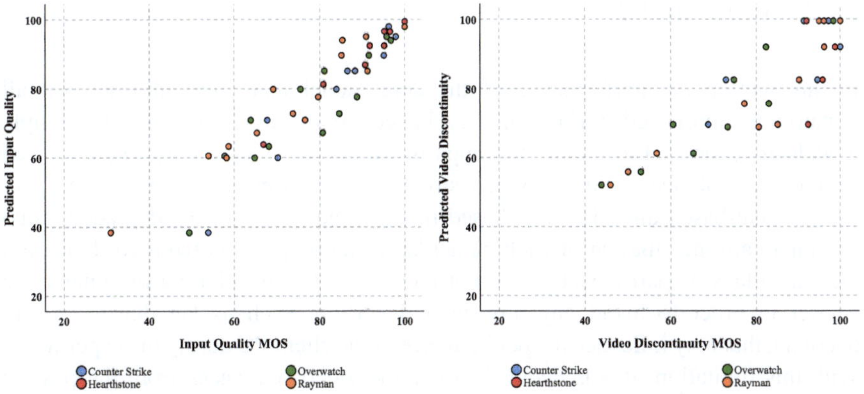

(a) (b)

Fig. 6.12 Scatter plots of predicted and assessed Input Quality and Video Discontinuity. (**a**) Predicted Input Quality vs. Input Quality MOS on validation interactive dataset. (**b**) Predicted Video Discontinuity vs. Video Discontinuity MOS on validation interactive dataset

Table 6.7 The performance of input quality and video discontinuity prediction models on the interactive dataset

	RMSE (R-scale)	RMSE (MOS-scale)	PLCC	SRCC
Input quality	6.07	0.25	0.92	0.93
Video discontinuity	9.63	0.39	0.85	0.86

prediction models (I_{trans_I} and I_{cntr_I}) are predicting the same assessed scale (input quality) in the subjective test, instead of separately evaluating them, the Eq. 5.9 that is presented in Chap. 5 is used to predict the input quality (O_{22}).

The performance of the input quality prediction model is evaluated based on the GIPS ratings, represent input quality as ground truth from the validation part of the interactive dataset. Figure 6.12a presents the scatter plot of predicted input quality and assessed input quality for the validation part of the interactive dataset. It can be seen that the model performs well in the prediction of input quality with a very low RMSE and high correlation as presented in Table 6.7.

The scatter plot shows that the model performs well on the validation dataset. Generally, the number of video games in the interactive dataset is lower compared to the passive dataset due to lengthy and expensive subjective experiments, which is one of the main limitations of this work. Thus, the model is only validated with four video games selected from different delay and frame loss sensitivity classes.

The impairment of video discontinuity can be predicted based on the Eq. 5.5 presented in Chap. 5. Based on the video discontinuity ratings collected in the interactive dataset and predicted video discontinuity, RMSE, PLCC, and SRCC are reported in Table 6.7 which show a high prediction performance for video discontinuity impairment. Figure 6.12b illustrates the scatter plot of the predicted video discontinuity and assessed video discontinuity in the validation dataset.

6.2.2 Core Model

In this section the performance of the core model that takes into account all impairments, proposed in Eq. 5.6, is evaluated. As discussed in Chap. 4, multiple models are proposed to predict the impairment due to video coding (I_{codV}) based on the level of access to the video stream information. To estimate the I_{codV} using signal-based and bitstream-based models, the access to the degraded video and bitstream information of each gameplay of participants is required. However, the gameplays of participants were not recorded when developing the interactive dataset in order to avoid any unwanted distortion, such as lag due to system overload, that may influence the performance of the client PC during the experiment. With this limitation, it was decided to use the video sequences from the passive dataset corresponding to the same game and encoding condition in the interactive dataset to measure the signal-based and bitstream-based models' prediction. Five video sequences are taken from CGVDS that cover five games in the interactive dataset, Bejeweled, Overwatch, Tekken, Rayman, and Worms. For the remaining four games, the video sequences are taken from the ITU-T Rec. G.1072 passive dataset. The ITU-T Rec. G.1072 passive dataset includes the CGVDS as well as another dataset named CGVDS Part2 (CGVDS-P2), which is built similar to CGVDS, including not only compression artifacts but also the effect of packet loss distortion.

Since the video quality of a single recorded gameplay is not equal to multiple gameplays of participants in the interactive dataset, the difference between the video quality MOS in the interactive and passive datasets for the corresponding stimulus is used for an adjustment. This difference, which describes the biases between the assessed video quality in passive and interactive test for a certain condition, named as $Scene_{bias}$, is predicted according to Eq. 6.2. Next, the $Scene_{bias}$ is added to the signal-based and bitstream-based model predictions to avoid errors due to the difference between the single recorded scene and played scenes. Thus, the video quality prediction (e.g., DEMI prediction) of a video sequence of the passive dataset, $\widehat{VQ}_{passive}$, is corrected according to $Scene_{bias}$, shown in Eq. 6.3.

$$Scene_{bias} = VQ_{interactive} - VQ_{passive} \tag{6.2}$$

$$\widehat{VQ}_{interactive} = \widehat{VQ}_{passive} + Scene_{bias} \tag{6.3}$$

Table 6.8 presents the performance of the gaming QoE model, considering different models predicting video coding impairment. The performance of the core model is not reported for NR-GVQM. NR-GVQM predicts the video quality in the VMAF range of score, and converting it back might add errors to the model. Thus, it is removed from the analysis of this section. Based on the table, the proposed gaming QoE model performs decently regardless of the selection of the model predicting video coding impairment. However, if the model has access to more information,

Table 6.8 Performance of the gaming QoE model according to the selection of the video coding impairment prediction model

	NDNetGaming	DEMI	BQGV	GamingPara	G.1072
PLCC	0.91	0.89	0.87	0.85	0.81
SRCC	0.92	0.91	0.89	0.89	0.86
RMSE	0.31	0.38	0.40	0.42	0.47

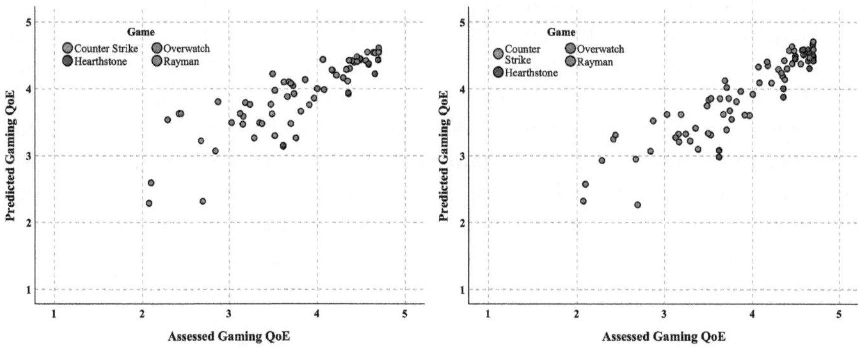

(a) (b)

Fig. 6.13 Scatter plots of predicted Gaming QoE vs assessed Mean Opinion. Score of Gaming QoE for validation part of interactive dataset. (**a**) Predicted Gaming QoE (based on GamingPara) vs. Gaming QoE MOS. (**b**) Predicted Gaming QoE (based on NDNetGaming) vs. Gaming QoE MOS

e.g., signal information of streamed video, it performs higher, as it can be seen for NDNetGaming.

It has to be noted that the validation dataset consists of video games from different classes of video complexity, as well as delay and frameloss sensitivity. In addition, an almost balanced number of distortion types, e.g., network and compression distortion, exist in the validation dataset. Therefore, even though the number of data points is not large, due to coverage of different distortion types and content classes, it gives an overall impression of how the model performs if employed in a cloud gaming service.

The model relies strongly on the classification of video games, and if the user of the model does not have knowledge about the game class, it has a significant impact reducing the performance down to $RMSE. = 0.78$ and $PLCC = 0.71$, if NDNetGaming is used as a choice of video coding impairment prediction model. Figure 6.13 presents the scatter plot of the predicted and assessed gaming QoE based on the validation interactive dataset based on GamingPara and NDNetGaming.

6.3 ITU-T Rec. G.1072

ITU-T Rec. G.1072 is developed based on large interactive and passive datasets. The interactive dataset and CGVDS are used partly in training and validation of the final model. While the model is in general developed similarly to the presented gaming QoE model, a few differences in the model design are discussed in Chap. 5. In addition to the differences in the model design, a few differences in training and validation of the model can be mentioned. First, as discussed earlier, a large passive dataset was used which include the CGVDS but also another similar dataset, CGVDS-P2 with more parameters such a packet loss. Next, due to the small number of ratings in an interactive test, it was decided to use a part of passive dataset in the validation of the model. Since the passive dataset has no ratings for overall gaming QoE, the overall gaming QoE was predicted based on the ratings from the video discontinuity and video quality that is fitted based on the interactive dataset after excluding conditions with the delay and packet loss, using a linear regression model as follows:

$$I_{QoE} = 1.6878 + 0.64218 \cdot I_{VQ} + 0.38857 \cdot I_{VD} \tag{6.4}$$

The I_{QoE} representing the delta-R of the overall gaming QoE was accurately fit with an RMSE of 3.56 on R-scale (delta R-Scale), Pearson correlation of 0.967, and adjusted r-squared of 0.932. The model is used to assign I_{QoE} for the passive dataset to be used as ground truth.

The model provides two modes: the extended mode that takes into account the video classification, and if the classification is not available to the model, the highest game class is assumed as a default mode. The performance of the core model is predicted according to the validation part of both interactive and passive datasets. The model reaches a high PLCC of 0.90 if the video classification is taken into account. If the default mode is considered, the model performance goes down to 0.82 in terms of PLCC (Table 6.9). The scatter plot of the G.1072 prediction and assessed MOS values are presented in Fig. 6.14. It can be seen from the figure, if the default mode is considered, the prediction values clearly deviate from the subjective ratings. This deviation from the subjective score could be stronger if more games from the low and medium classes are considered.

Table 6.9 Scatter plot of the G.1072 predicted and assessed MOS ratings on the test dataset without using content classification (default mode) on the left, and using the content classification (extended mode) on the right

	Without considering classification (default mode)		Considering classification (extended mode)	
	R-scale	MOS-scale	R-scale	MOS-scale
RMSE	12.19	0.47	8.03	0.33
PLCC	0.80	0.82	0.89	0.90

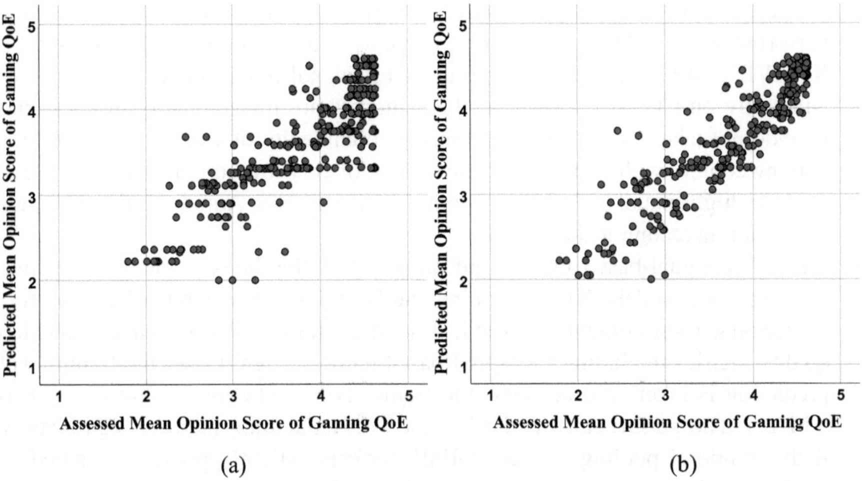

Fig. 6.14 Scatter plot of the predicted and assessed MOS ratings on the test dataset without using content classification (default mode) on the left, and using the content classification (extended mode) on the right, based on ITU-T Rec. G.1072, Figure G.1072(20)_F02 in [51]. (**a**) Default mode. (**b**) Extended mode

6.4 Summary

In this chapter, the performance of the proposed impairment prediction models is reported on validation dataset(s). The first section is dedicated to the performance evaluation of video coding impairment prediction models. These models that are known as video quality models are compared to the SoA video quality models. The important findings of this section are summarized as follows:

- Among evaluated planning models, the proposed GamingPara performs best in terms of PLCC and SRCC. The coding impairment of ITU-T Rec. G.1072 also performed very well on all three datasets. It has to be noted that both mentioned models rely on having access to the classification of video games. Therefore, the performance is expected to decrease if no knowledge about the game video complexity is available or the classification fails in correct assignment of the classes.
- With respect to the performance of evaluated bitstream-based models, two ITU-T Recommendations, P.1203 Mode 3 and P.1204.3 (FHD mapped) correlate very well with the assessed video quality, considering the fact that they were initially not developed for gaming video streaming applications. The proposed BQGV as a lightweight model performs best among the evaluated bitstream-based models on CGVDS. However, it has to be noted that this model is validated using leave-one-out cross-validation (one refers to one bin consisting of 3 source video sequences), and it might be biased to gaming content as well as the special encoding setting used in CGVDS.

- The evaluation of FR signal-based models on CGVDS revealed the reliable performance of VMAF for gaming content, which was previously shown on KUGVD and GVSET. Other traditional FR signal-based models perform satisfactorily on CGVDS. However, these models are image quality metrics, and they do not take into account the temporal complexity of videos. Therefore, for gaming content with scenes that have a range of very low complexity, e.g., Tetris, to a very high spatial and temporal complexity, e.g., Nier Automata, these models might not be recommended to use.
- Three NR signal-based models are proposed in this book. Among them, two deep learning models, NDNetGaming and DEMI, showing a remarkable performance on gaming content. NDNetGaming outperforms all evaluated image/video quality metrics, including FR signal-based models, even if only the frame-level prediction is taken into account (the frame-level prediction is almost equal to VMAF). The performance of DEMI and NDNetGaming increase significantly if the temporal pooling is used. DEMI performs slightly poorer compared to NDNetGaming on the gaming dataset, whereas it outperforms NDNetGaming on a non-gaming dataset. The proposed lightweight NR model, NR-GQVM, shows a satisfactory result on all three gaming video quality datasets. Considering the model's simple training process, it can be easily extended for new types of content, video codec, or type of distortion.

Next, the performance of other impairment prediction models of the gaming QoE model is evaluated. Two impairments of control latency (i_{cntr_I}) and video transmission error (i_{trans_I}) are evaluated together as they are assessed using the same scale in the interactive test, input quality (GIPS scale). The result on the validation part of the interactive dataset shows a high PLCC of 0.92 to predict the input quality. This promising result is achieved if the gaming classification towards delay and frameloss sensitivity is considered. Therefore, the performance could decrease significantly if the model does not take into account the gaming classification. The effect of video transmission error on video quality is modeled based on the video discontinuity item. The model obtains a satisfactory result with a PLCC of 0.85.

Finally, the proposed gaming QoE model is evaluated based on the selection of different video coding impairment models. The result suggests a high performance of the model regardless of the video coding impairment model's choice. However, if the model has access to more video stream information, e.g., bitstream or signal information, the prediction performance increases considerably. It has to be noted that, for the evaluation of signal-based and bitstream-based models, the representative video sequences of each game are used, taken from the passive datasets.

In addition, the performance evaluation of ITU-T Rec. G.1072 is presented in the last section based on reported PLCC and RMSE in the recommendation. The performance of ITU-T Rec. G.1072 is evaluated based on the interactive and passive dataset. The result revealed the importance of gaming classification in the improvement of final prediction, from PLCC of 0.82 if the default mode is considered to 0.90 if the extended mode (gaming classification) is considered.

Chapter 7
Future Challenges: Enhancement Techniques

7.1 Super-Resolution Models

In the past few years, several works have been published aiming at development of enhancement techniques that could up scale images and videos to a higher resolution without significant quality loss, known as the Super-Resolution (SR) task. Recently, there has been a huge improvement in the performance of the methods for SR task, primarily due to two big factors: advancement in deep learning methods (such as deep CNN), and improvement in loss function to evaluate the quality of the enhanced image. Several new CNN architectures have been introduced that improve the prediction of different tasks by either proposing a deeper network (e.g., VGG [121]) or enhancing the flow of information (e.g., DenseNet [100] and ResNet [101]). Besides, new and more versatile machine learning methods such as Generative Adversarial Network (GAN)s [122] and Auto-encoders have found their application in SR tasks [123]. GANs, which have two main blocks, generator and discriminator has recently gained lots of attention for quality enhancement, especially for SR tasks. In SR tasks, the generator is responsible for generating enhanced images based on the received distorted input, while discriminator evaluates the generated image quality compared to the real image. GANs have been considered as a solution for quality enhancement of distorted images due to compression, noise, or block loss [124].

In this section, the development of a super-resolution model is described, and it will be discussed how these models could offer new challenges for quality assessment. In addition, some of the gaming quality metrics introduced in this book are used to evaluate the developed model. This section is mainly written based on the data and text from the author's publication [119] with some new measurements.

S. Zadtootaghaj, *Quality of Experience Modeling for Cloud Gaming Services*,
T-Labs Series in Telecommunication Services,
https://doi.org/10.1007/978-3-030-98249-2_7

7.1.1 *Super-Resolution Quality Assessment*

One of the well-known competitions for the development of quality enhancement techniques is the grand challenge at *Perceptual Image Restoration and Manipulation (PIRM)* workshop of ECCV 2018 [123] which aims to investigate models for SR tasks using different types of models and loss functions. The SR models can be categorized differently based on different criteria. For instance, they may use a different network architecture, where older models focus on the CNN core models, the new version takes advantage of GAN architect or image transformers. They could also be categorized based on the target signal; some focus on image content, and others on video by taking the temporal aspect into account. However, one of the most important aspects, which is highlighted in [123] is the aim of the models to improve quality. Some models target reconstruction accuracy which the primary purpose is to reconstruct the image as close as possible to the reference image. Therefore, these models typically result in an improvement of full-reference objective metrics, such as PSNR and SSIM. The second category aims to improve the perceptual quality, which does not necessarily improve the reconstruction quality; consequently, the enhanced image/video may receive an excellent quality judgment from the human eye but a lower score in terms of a full-reference model. In this work, the reconstruction accuracy is measured based on full-reference metrics, while the perceptual accuracy is measured based on a no-reference quality metric and subjective ratings.

7.1.2 *Dataset for Training*

In order to dive into the performance of signal-based models on enhanced gaming videos, a gaming specific enhancement model is developed. In this subsection, the dataset for training and evaluation is discussed.

Two subsets of low and high-quality sets are created from extracted frames of multiple encoded sequences of the game League of Legends (LoL) to build the training dataset. For the lower quality frames for the training set, the frames are extracted from encoded gaming video(s) compressed using the H.264/AVC compression standard using the FFmpeg libx264 encoder wrapper. The videos were encoded in multiple resolution-bitrate pairs (bitrates ranging from 300 kbps to 2000 kbps and three different resolutions, 480p, 720p, and 1080p). All the 480p and 720p encoded video sequences were up-scaled to 1080p using the *bilinear* scaling function of FFmpeg tool. In addition, the corresponding reference-quality frames are also extracted from uncompressed recorded raw videos. Hence, the training set consists of two sets of high-quality reference frames as well as lower quality compressed and/or up-scaled frames. For the training dataset, a total of 11k frames are extracted (pairs of high and low-quality frames) from the videos randomly, and each is cropped to the size of $96 \times 96 \times 3$. For more details on the patch selection, the reader is referred to the reference publication [119].

7.1.3 Model Development

In this subsection, the GAN architecture that is used for video gaming enhancement is introduced. The model uses the state-of-the-art GAN model, SRGAN [125], as the starting baseline model, which is then further modified to improve the efficiency. Generally the following modifications are applied to the original SRGAN:

- **Loss Function:** SRGAN uses a loss function, which is a weighted sum of two components, content loss, and adversarial loss. The content loss was defined to replace the pixel-based approach with a perceptual metric, which allows the model to increase the perceptual quality instead of reconstruction accuracy. SRGAN measures the content loss based on the feature representations from the pre-trained VGG19 model. Thus, the euclidean distance between the feature representations of a generated image and the reference image defines the content loss. The adversarial loss is then determined by the probability based on the discriminator allowing to differentiate between the generated patch and reference patch. To further improve the loss function, the VGG19 is retrained for gaming image quality tasks to improve the model performance. Hence, another loss function is added to the model, called distortion loss, which is based on retrained VGG19 network for the image quality prediction task of gaming content. To retrain the VGG19, the frame-level VMAF values are used as the ground truth. The final loss function is a weighted average of SRGAN loss functions and retrained VGG19 loss (distortion loss).
- **Network Architecture:** For the generator block, the residual-based network in SRGAN is replaced with U-Net [126] network that can improve the flow of information as well as lead us to higher texture reconstruction. In addition, the Parametric Rectified Linear Unit (PReLU) function [127] is used instead of the commonly used ReLU or LeakyReLU, as it has been shown that this activation function works better for compression related tasks [128].

7.1.4 Objective and Subjective Measurement

In this subsection, the performance of different signal-based quality models on the enhanced images from the developed model is investigated. It is important to understand if the current quality metrics are efficient enough to take into account the adversarial distortion from the model. To quantify the quality enhancement, both NR and FR quality metrics are considered. In addition, a small-scale image quality subjective test is conducted to investigate the performance of each model.

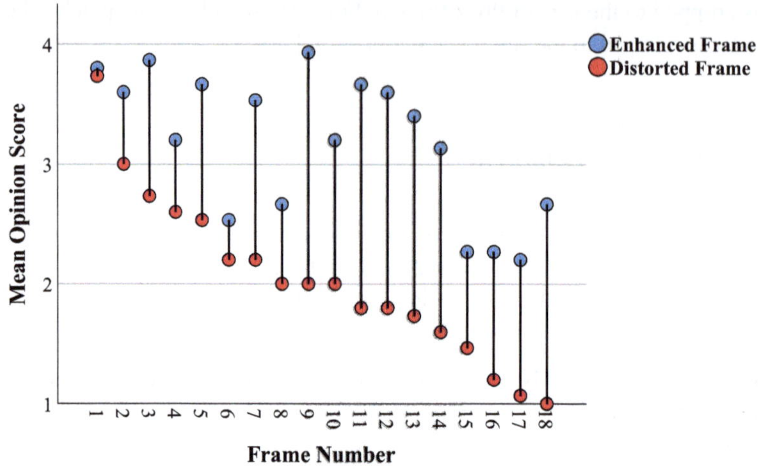

Fig. 7.1 Mean opinion score difference between enhanced and distorted frames taken from [119]

In order to quantify the perceptual enhancement of the super-resolution model, a small scale subjective test conducted with 15 subjects. For the image subjective test, two frames from the game LoL were selected and used. Eight distorted frames are selected from each reference frame which have different types and levels of distortion. In total, 38 images including 18 distorted images, 18 enhanced images and 2 reference images are used in a the subjective test adhering to ITU-T Recommendation P.910 [3]. In addition to the image quality rating, the participants were asked to rate the level of blurriness and blockiness in the image on a 5-point ACR scale. This allows to investigate the performance of the model DEMI which is capable of predicting the quality of blurriness and blockiness artefact separately. Figure 7.1 plots the difference of MOS between the distorted and corresponding enhanced frames in order to show the perceptual enhancement power of the proposed method. In Fig. 7.1, a maximum 2 MOS difference can be observed (in frame 9, affected by blur).

The enhancement in terms of MOS for frames with blockiness (frame 1, 2, 4, 6, 8, 10, 15, 16, 17) is lower than that observed for the frames with blurriness artifact (rest of the frames). The reason behind this can be attributed to the fact that blurriness is more uniformly distributed and hence easier to learn and also to the fact that the baseline SRGAN model is designed to minimise such artifact. The reader can refer to the reference publication [119], for more information about the performance of the proposed GAN model.

As a choice of quality metric, the NR metrics, including PIQE [114], and NIQE [94], DEMI, and NDNetGaming are selected. In addition, two FR quality metrics of PSNR and SSIM are also used. For evaluating the NDNetGaming metric, only the image quality prediction (output of the second step) is used since the evaluation dataset is image content. Similarly, for DEMI, only the output of Fragmentation and

Table 7.1 Performance of signal-based models on evaluation dataset

Signal-based models							
Metrics	PSNR	SSIM	PIQE	NIQE	NDG[a]	DEMI Blur	DEMI blockiness
PLCC	0.57	0.54	−0.64	−0.77	0.90	0.80	0.75
SRCC	0.51	0.53	−0.62	−0.79	0.89	0.78	0.73

[a] NDG stands for NDNetGaming

Unclearness models are used and compared to blockiness and blurriness ratings. Other signal-based models such as VMAF and NR-GVQM are not used as they require information from the temporal domain as input to the model.

Table 7.1 presents the performance of evaluated signal-based models on the evaluation dataset. As it can be observed, the FR metrics did not perform well. This might be due to the low reconstruction accuracy of the GAN model for enhancement. Among the NR metrics, the NDNetGaming outperforms all other models and clearly shows that the model can be used for the assessment of the perceptual quality of super-resolution models. It has to be noted that the DEMI was only evaluated with blurriness and blockiness ratings from the dataset. DEMI's sub-components do not perform well to predict blurriness and blockiness ratings. Similarly, the two other NR models, PIQE and NIQE, failed to perform well on the gaming dataset.

7.2 LCEVC

While super-resolution models are one of the kinds of enhancement models expected to find their way to the video gaming streaming market, a standard codec has recently been designed for video quality enhancement. In October 2020, MPEG finalized the MPEG-5 Part 2 Low Complexity Enhancement Video Coding (LCEVC) standard that improves the coding efficiency [129].

LCEVC is a hybrid codec that has two enhancement layers that enhances the encoded stream, which applies on top of a base codec. The base codec potentially could be H.264, H.265, VP9, or AV1. The enhancement layers provide features to the base codecs to improve the details and sharpness of the final decoded sequence and reduce the encoding/decoding complexity.

LCEVC has been shown to improve the quality of a wide range of video content [129] including gaming videos [120]. In this section, the performance of a few selected signal-based video quality models are investigated to see if the current FR and NR models are capable of accurate measurement for video sequence enhanced by LCEVC. For the sake of brevity, only two well-known signal-based models, VMAF and PSNR, as well as two gaming video quality models, NDNetGaming and DEMI, are investigated in this section. For more information about the enhancement offered by LCEVC, the reader can refer to [120].

Table 7.2 Summary of video encoding parameters

Parameter	Value
Duration	30 s
Resolution	1080p
Bitrate (kbps)	800, 1500, 3000, 4000, 6000
Framerate	60 fps
Encoder	FFmpeg
Encoding Mode	CBR
Video Compression, Standard	H.264, LCEVC H.264, H.265, LCEVC H.265
Preset	Medium and veryfast

7.2.1 Subjective Test Design

For the subjective experiment, five video sequences, BlackDesert, Dauntless, GTA5, Overwatch, and Tekken, were selected because of their high spatial and temporal complexity. The reference video sequences are taken from the CGVDS dataset, and they are encoded using the SDK provided by V-Nova, including FFmpeg with LCEVC support for x264 and x265 as base codecs, referred to as LCEVC-x264 and LCEVC-x265, respectively. The reference video sequences are encoded at 1080p resolution and CBR rate controller mode. All video sequences are encoded in multiple bitrate pairs using native-x264, LCEVC-x264, x265, and LCEVC-x265. For encoding, the medium preset is selected for x264 and LCEVC-x264, and veryfast preset is selected for x265 and LCEVC-x265. Table 7.2 summarizes the encoding parameters.

The subjective test is designed for two purposes, to compare the enhancement that LCEVC offers compared to the native codec (which is presented in [120]) and to measure the quality of decoded video sequences. Therefore, a special test designed is the approach. The subjective test was conducted using a Pairwise Comparison (PC) using a 7-point extended continuous (EC) Absolute Category Rating (ACR) rating scale, following ITU-T Rec. P.809.

During the experiment, a pair of videos from the same codec family (e.g., x264 vs. LCEVC-x264 and x265 vs. LCEVC-x265) were shown sequentially (in random order), and subjects were asked to rate both of them using two separate 7-point scales provided after watching both sequences. In total, 18 ratings were collected were ratings from two participants were removed due to extreme outliers. Similar instruction and pre-questionnaire that were used to develop the CGVDS dataset are used in this work.

Table 7.3 Performance of signal-based models on LCEVC test dataset

	PLCC				SRCC			
	PSNR	VMAF	NDG[a]	DEMI	PSNR	VMAF	NDG[a]	DEMI
LCEVC-x264	0.71	0.93	0.92	0.91	0.71	0.93	0.92	0.91
LCEVC-x265	0.68	0.92	0.92	0.90	0.69	0.93	0.95	0.92
Native-x264	0.80	0.96	0.97	0.95	0.82	0.96	0.98	0.93
Native-x265	0.78	0.92	0.95	0.94	0.79	0.93	0.94	0.91
LCEVC	0.70	0.92	0.92	0.91	0.69	0.93	0.93	0.91
Native	0.79	0.95	0.95	0.94	0.81	0.95	0.95	0.92
All data	0.75	0.92	0.89	0.87	0.76	0.92	0.85	0.84

[a] NDG stands for NDNetGaming

7.2.2 Results

The performance of four signal-based video quality metrics, PSNR, VMAF, NDNetGaming, and DEMI, for the LCEVC codec is investigated by measuring the correlation of each metric, presented in Table 7.3. Thus, the PLCC and SRCC of these four metrics are measured separately for LCEVC and each base codec, Native codecs, all LCEVC data, and all Native data separately. The correlation analysis reveals that VMAF, NDNetGaming, and DEMI perform extremely well across the dataset regardless of codec usage. However, all metrics performs higher for the native codecs compared to LCEVC. Even PSNR seems to have relatively high performance in general, but there is a 0.9 point difference in terms of PLCC between LCEVC and Native codes. Therefore, it can be concluded that LCEVC improves the perceptual quality as well as reconstruction quality. Based on the correlation results, NDNetGaming and VMAF perform similarly. Considering the fact that NDNetGaming is an NR metric compared to VMAF as an FR metric, it shows that NDNetGaming is a pretty robust metric for quality assessment of gaming content with codecs other than H.264, which was tested in the previous chapter.

7.3 Summary

In this chapter, the video quality enhancement techniques are discussed, which is expected to influence the future of gaming streaming services. Therefore, two different enhancement approaches, a super-resolution model and LCEVC standard codec, that have been proven to enhance gaming videos are investigated in this chapter.

The investigation focused on the performance of a few signal-based models to understand if there is a need to develop new metrics for enhanced gaming videos. It must be noted that planning and bitstream-based models are not investigated as super-resolution enhancement models are applied after decoding the signal.

For LCEVC, the bitstream data format have a complete different structure due to enhancement layers, and it is clear that the current developed model would not work.

Among signal-based models introduced in this book, the performance of two deep learning-based video quality models, NDNetGaming and DEMI, are investigated for these enhancement techniques.

The results showed that the FR metrics fail to accurately predict the super-resolution models that target to increase perceptual quality and not reconstruction quality which is in line with observation in [123]. Surprisingly, the widely used NR metrics, PIQE and NIQE, in this field did not correlate well with subjective data. In contrast, NDNetGaming performs well on the validation dataset.

The results for LCEVC showed that in contrast to super-resolution models, the FR metrics could be used for the measurement of the quality that this codec improves. This is very important as typically, the codec comparison is conducted using rate-distortion curves which an FR metric such as PSNR or VMAF is used. In addition, NDNetGaming performs as well as VMAF which reveals the high performance of this model as an NR metric on gaming content.

Chapter 8
Conclusion and Outlook

8.1 Summary

The present book aims at the development of gaming QoE model that can be used for different purposes, including, but not limited to, network planning, quality monitoring, resource allocation, and benchmarking cloud gaming services. The scope of the model is presented in Chap. 1 that defines the considered influencing factors, range of parameters, participants characteristics, end-user devices, targeted types of models, and modes of operation.

Chapter 2 starts with an introduction to Quality and Quality of Experience as fundamental concepts. A cloud gaming taxonomy developed in 2013 is described in this chapter, which overviews the important gaming QoE influencing factors and quality features that could potentially be considered for developing a gaming QoE model. In collaborative work with other researchers, the author contributed to the development of ITU-T Rec. G.1032 [1] that identifies a long list of gaming QoE influencing factors in the context of cloud gaming services that is summarized in the second chapter of the book. As an imperative prerequisite before the development of a QoE model, a standardized subjective test methodology is required. Therefore, in this book, ITU-T Rec. P.809 is followed. This Recommendation defines two types of paradigms for quality assessment of cloud gaming services; first, passive viewing-and-listening tests targeting the output quality of the service, and, second, interactive tests that aim to assess other quality features influencing the gaming QoE of cloud gaming services. In addition, two questionnaires are described to assess the interaction quality and video quality sub-dimensions. These two questionnaires are later used in the development of gaming quality datasets. Next, an overview of efforts towards the development of gaming quality models is given. While several studies can be found in the literature, they are all limited to a small number of parameters and targeting only a specific quality features. The studies were conducted in different laboratories under different subjective test methodologies

S. Zadtootaghaj, *Quality of Experience Modeling for Cloud Gaming Services*,
T-Labs Series in Telecommunication Services,
https://doi.org/10.1007/978-3-030-98249-2_8

that cannot be generalized. Finally, the chapter ends with a discussion on the performance of existing video quality models on gaming video content. Two studies are described that the author contributed to, which evaluate the performance of multiple well-known signal-based video/image quality metrics on gaming video content. Studies suggest that there is a need for the development of NR video quality models for gaming content.

Chapter 3 describes the necessary steps to develop a gaming QoE model. First, the information about the structure of the model framework is provided. The model follows a modular structure that is based on separated impairment-factors that influence the gaming QoE. Three impairment factors of video coding, transmission error, and control latency impairment are defined. Such a modular structure offers flexibility in the development of separate models for each impairment factor. In addition, the model could get updated easier and faster for potential extensions. For the development of the model, it is decided to develop multiple models for video coding impairment factors based on the viewing-and-listening tests. This allows the use of the model in different monitoring points of the network depending on the level of access to the video stream. Also, multiple datasets are created to train different gaming QoE models, including two gaming video quality datasets, an image quality dataset, and a large interactive dataset following test methodology recommended by ITU-T Rec. P.809. The chapter ends with the development of three video classifications that are necessary for gaming QoE model development. The proposed game classifications provide information to label the games based on their sensitivity towards delay and frameloss as well as video coding complexity. The developed classifications are used later on for the development of the quality models for cloud gaming services.

Chapter 4 presents the models that are developed to predict the coding impairment factor. The models are developed based on gaming video quality datasets for different purposes. First, three planning models are described that can predict the gaming video quality according to simple encoding parameters. Among them, GamingPara is developed by the author, which follows the multidimensional approach (DBSQE-V) to predict the video quality according to its sub-dimensions of fragmentation, unclearness, and discontinuity. Such a multidimensional approach allows a more in-depth insight into the causes of low video quality, known as diagnostics information. Next, three bitstream-based video quality models are described. This type of model is a useful tool to monitor the video quality at different network monitoring points. Depending on the level of access to the bitstream information, they can be divided into payload-based and header-based models. While the payload-based models are typically more accurate, the latter is lightweight and much simpler to deploy in the network. The author developed the BQGV model that extracts the packet header information to predict video quality and its sub-dimensions. In addition, two well-known bitstream-based models developed under ITU-T SG12 are described in this chapter. Finally, three NR signal-based video quality models are proposed for gaming content. As discussed earlier, based on the previous studies, it has been shown that FR metrics perform well on gaming content. However, the SoA NR metrics fail to predict the video quality of gaming content accurately. Thus, the main focus is devoted to the development of NR metrics for gaming videos. First, a lightweight model, NR-GVQM, is developed based on low-level image features and targeting an objective metric, VMAF, as ground truth. The idea was to develop a model to fill the gap between FR and NR models' performance without having access to a large subjective dataset. The model can be easily extended with a larger dataset of recorded gameplay, a new type of codec, and trained based on other high-performance video quality models. Besides, two deep learning based video quality models are developed. NDNetGaming and DEMI are both trained based on three phases of VMAF training, fine-tuning based on subjective ratings, and temporal pooling. However, there are some fundamental differences in the design and training of the models. In the development of the NDNetGaming, it is carefully investigated which well-known CNN architecture is a good candidate for the image quality estimation task based on the two criteria of high performance and a small number of trainable parameters. Also, the NDNetGaming uses a simple temporal pooling method that does not add too much complexity to the model. DEMI is proposed to improve the limitations that are not considered in NDNetGaming, including considering both gaming and non-gaming content in the training phase, multi-scaling feature extraction, and usage of a more complex but accurate temporal pooling method. In addition, DEMI provides two diagnostic frame-level scores for fragmentation and unclearness.

Chapter 5 provides three models to predict the remaining impairment factors, impairments due to transmission error, and control latency impairment. The earlier does impact both video quality and input quality, while the latter only affects the input quality. The impairments due to transmission error are modeled based on the frameloss rate and encoding framerate, once using discontinuity ratings as target labels, and once using averaged GIPS items indicating the input quality. The control latency impairment is modeled based on the network delay as input and ratings for GIPS items as ground truth. All three impairment models are trained based on the interactive dataset presented in Sect. 3.4. Finally, the core gaming QoE model is presented, which is developed upon the model framework structure presented in Chap. 3. In addition, the ITU-T Rec. G.1072 model is described in detail and the differences compared to the model framework of the book are explained.

Chapter 6 evaluates the performance of proposed models in terms of PLCC, SRCC and RMSE. First the performance of the video coding impairment models are presented. The performance is evaluated on the 5-point MOS scale to compare it with existing video/image quality models. In general the main findings can be summarized as follows:

- The proposed GamingPara outperforms the ITU-T planning models in terms of PLCC and SRCC. However, it underestimates the quality for datasets created with software encoders, GVSET, and KUGVD, leading to higher RMSE than other models.
- Among bitstream-based models, the proposed BQGV as a packet header-based model performs similarly to existing payload-based ITU-T models, P.1203 mode

3 and P.1204.3 on CGVDS dataset. However, it performs poorly on datasets that are created with different encoding presets.
- The proposed NR model, NDNetGaming, outperforms all well-known video/image quality metrics, including FR and NR metrics, across all three gaming video datasets. However, it does not perform similarly well on non-gaming content.
- DEMI frame-level prediction performs similar to NDNetGaming on gaming content while performing quite well on non-gaming content.
- If the temporal pooling of NDNetGaming and DEMI is applied, they both perform significantly better compared to all video/image quality models across all gaming dataset, and similarly as well as VMAF on non-gaming datasets.

In addition to the coding impairment, the performance of other impairment models is investigated in Chap. 6. The results show a high performance of impairments models that predict the interaction quality, based on I_{trans_I} and I_{cntr_I}, with PLCC over 0.90 on both training and validating datasets. The performance of the proposed model to predict video discontinuity due to transmission errors (I_{trans_V}) is reasonable, PLCC of 0.85 on full dataset, considering the small number of conditions in the dataset triggered by transmission errors.

The core model performance is evaluated on the validation part of the interactive dataset. Depending on the selection of the coding impairment prediction model, the performance could vary, in terms of PLCC, from 0.80 to 0.90. The best performance is achieved if signal information is available and if the NDNetGaming is selected as a coding impairment prediction model. It has to be noted that for an accurate prediction, the model relies strongly on the information about the game class. If the model does not have access to the information about the game classification, the high class of complexity would be selected, which significantly decreases the performance, from PLCC of 0.90 to 0.71 if NDNetGaming is used for I_{cod_V}.

Chapter 7 evaluates the performance video quality metrics on the two enhancement approaches applied to gaming videos streams. It was shown that the FR metrics might not be an excellent option to be used for the quality measurement of super-resolution models. In addition, the results of this chapter reveal that NDNetGaming performs robust across different codecs and enhancement techniques.

8.2 Contributions of Book

The book's main contribution is the development of a gaming QoE model based on relevant impairment factors to estimate the quality experienced by players of cloud gaming services. The model uses a series of video coding impairment prediction models that allow using the framework for different purposes, e.g., monitoring the quality, and based on the different levels of access to video stream. In addition, the following contributions of book can be mentioned:

- Identifying quality factors influencing the quality of cloud gaming services.

- Creation of subjective test databases of video quality and gaming QoE using the passive and interactive test paradigm, which allow the development of QoE models
- Development of a video game classification that classifies games according to their sensitivity towards delay, frameloss, and video complexity.
- Development of multiple models for predicting the video quality of gaming content based on the level of access to information about the degraded video, e.g., bitstream information and signal information.
- Contribution to the three ITU-T Recommendations that are described in the present book, P.809, G.1032, and G.1072.
- Contribute to open-source projects by providing the source codes for multiple developed models, NDNetGaming, GamingPara, DEMI, and ITU-T Rec. G.1072, as well as making the video and image quality datasets publicly available. This allows other researchers to develop and improve the gaming QoE model for cloud gaming services in the future.

8.3 Limitation

Cloud gaming is one of the most challenging multimedia services that run on top of IP-based networks. The development of a holistic quality model for cloud gaming services is very challenging and required the author to limit the scope of the book to more realistic development of models that considers the interest of stakeholders, cloud gaming and network providers, as well as a limited number of quality factors and influencing factors. In the following, the limitation of the book will be discussed.

Users: Within the scope of the book, it is decided to target casual players who have a fair experience and interest in playing games. Core gamers are not considered for multiple reasons. Core gamers are very sensitive towards degradations, even small degradations, e.g., 100 ms delay, could be blamed for the bad experience such as loss in a game and cause the departure of players of the service. This can be seen from statistical reports [130] suggesting that most of the cloud gaming service users are casual gamers and not core gamers. Also, unlike the other services, several core gamers typically play games with high-end technologies such as 144 Hz and High-Dynamic-Range (HDR) displays. Thus, they may rate the reference condition in a subjective test much lower compared to other players, just because of having different expectation. This was also observed in the interactive test, which results in labeling some of the participants as experienced players. However, with the advancement of network and multimedia technologies, it is expected that more experienced gamers join the cloud gaming service; thus, they should not be neglected from the development of gaming QoE models.

Social Context Factor: It has been shown that players do not only play the games to have fun but also to communicate with other players in the context of multiplayer gaming. However, measuring the influence of social context on gaming QoE is a challenging task. Hence, following ITU-T Rec. P.809, this factor is not considered in the subjective test, while the impact of this factor remains unclear.

Quality Features: In this work, only input quality and output quality are considered as the most important quality features. The structural equation modeling conducted by Schmidt [48] on the interactive dataset shows that 62% of the variability in the gaming QoE ratings can be explained only by the individual ratings of input quality and output quality. If other quality features related to player experience are taken into account, only 9% higher variability in the gaming QoE ratings (71%) can be explained by the model. However, games with certain design flaws might influence the importance of aspects such as appeal and flow. Besides, if the study is conducted in a longer stimulus duration, this may influence the importance of aspects such as immersion.

Subjective Test: Within the scope of this book, the studies are designed following ITU-T Rec. P.809. As a choice of the questionnaire, a total of 31 items are selected as the post-condition questionnaire. Such a long post-condition questionnaire might introduce fatigue during the subjective test, which influences the ratings of participants, especially for the latter items. For the future subjective test, the iGEQ items could be removed to reduce the number of items if only output quality and interaction quality are considered for the development of the model. All studies are conducted in short stimulus duration, as a consequence long-term effects are not considered in this work.

Video Games: The video games and the scenarios for the subjective test were selected according to some criteria to ensure that different players perceive a very similar experience and to reduce the influence of player performance or strategy in the game. Therefore, video games with multiple mechanics or requiring participants to be involved to the story of the game are not offered to players. In addition, this work is limited to the number of video games that are selected for the test, which might not include the full spectrum of genres and game mechanics.

Game Classification: The game classification presented in Sect. 3.6 is a great help to develop accurate gaming QoE models. However, the performance of gaming QoE models depends strongly on the performance of these classification methods, which still rely on human judgment to identify the game characteristics.

System Technology: The cloud gaming service is still on the rise, and new technologies rapidly evolve the experience of the players. For example, recently advanced video codecs, such as H.266/MPEG-I Part 3 (known as VVC) and AV1, offer better quality at the same bitrate level compared the H.264/MPEG-4 AVC that is used in this book. However, due to additional latency that might add to the service, they are not offered by any well-known cloud gaming service. While it could be assumed that the proposed signal based models might be still valid, this needs to be investigated. In addition to the advancement of video coding techniques, the advancement of output modalities such as VR HMD and HDR displays (which require HDR content) are also new challenges that are required to be addressed in the near future.

System Parameters and Range of Parameters: In this book, the parameters and range of parameters are limited to allow training the model within a reasonable number of subjective tests. In the following, a short list of essential parameters and the range of parameters that might be of interest for future work is provided.

- *Jitter* is one of the important parameters that is neglected in this work. Delay and Jitter are innately linked, and for selecting the range of Jitter, it must be considered in association with delay.
- *Network packet loss* in this book is only applied following a uniform pattern, while for more realistic modeling, complex patterns could be considered as well.
- *Video Coding Setting* follows a typical setting considered according to the NVIDIA cloud gaming service. However, this setting might not be the same for other cloud gaming services.
- *Audio Coding* parameters are not considered since the audio signal is relatively smaller compared to the video signal and probably will be less affected by the network and encoding degradation. However, transmission error on audio signal and audio coding might influence the player experience, which is out of the scope of this work.
- *Short range delay* might be of interest if core gamers are considered in the scope of the model. Based on the collected data through the interactive test, it can be observed for many games that participants cannot distinguish the quality between a reference condition and a condition with a delay under 100 ms. This might not hold true if the core gamers are the target group. Thus, the selection of more delay values below 100 ms might be of interest for accurate quality prediction.

Data Collection: Conducting a subjective test is an expensive process, especially considering the lengthy interactive paradigm. This limitation does not only affect the number of testing parameters and a selected range of parameters but also affects the size of the dataset for training and validating the models. Consequently, the training and validation datasets are split according to the game class labels to ensure a balanced number of each game class in both training and validation datasets. However, this human biased splitting of the dataset to validation and training sets might positively influence the performance of the models. Also, the validation dataset might not be large enough to draw a confident conclusion about the performance of the developed gaming QoE model.

Performance Evaluation: The MOS values always have a statistical uncertainty, which is commonly explained by the confidence interval. This uncertainty potentially affects the correctness of the results reported using ranking order metrics and error measurement metrics. ITU-T Rec. P.1401 [117] recommends using a statistical metric called epsilon insensitive RMSE (known as RMSE*), when the uncertainty of the subjective scores is taken into account. This RMSE considers the 95 % confidence interval of the individual MOS scores. Similarly, Naderi et al. [131] suggest a transformation to MOS values before measurement of ranking-based correlation, e.g., SRCC. However, Naderi showed that the effect of uncertainty to SRCC measurement approaches to zero when the number of conditions in the dataset increases. In addition, if the number of ratings per condition is high, typically the confidence interval becomes smaller. A smaller confidence interval causes RMSE* gets closer to the absolute RMSE. Therefore, due to the high number of conditions as well as a sufficient number of ratings per condition in all datasets used in the performance analysis of the models, such a MOS transformation and RMSE* are not applied in this book. However, the effect of subjective uncertainty remained unclear in this work.

Considering these limitations, this work is the most extensive work conducted in the research community to predict gaming QoE. However, cloud gaming is rapidly integrating new technologies into the system, which requires extending the model for upcoming changes. In the next section, the important aspects that must be considered for extending the current work are discussed.

8.4 Model Extensions

The proposed model framework in the presented book offers flexibility to extend the model for recent technologies adapted to the cloud gaming service, such as recent video codecs, HDR, and VR technologies. This is possible with the modular structure that allows updating each impairment factor independent from others. For example, in a possible extension of the model for a new encoding setting or codec, only I_{codV} requires to be retrained, and the prediction of other impairment factors should still hold valid.

In the following, a short list of possible extensions of models is listed, which at this point of the time can be foreseen as valuable and necessary steps towards developing a holistic gaming QoE model.

- *Advanced Codecs*: It is expected that new codecs soon be integrated into cloud gaming services. As an example, Google announced that they are working on using the AV1 codec in their cloud gaming Stadia service.[1] To extend the developed model for a new codec, the I_{codV} must be considered to be retrained. While it is expected that the proposed signal-based models still perform decently, the bitstream-based and planning are required to be adapted.
- *Bitstream-based Video Quality Models*: In this book, a bitstream-based model is proposed that uses the packet header information to predict the video quality. This type of model is considered to be lightweight, accurate, and easy to be employed on network monitoring points. Thus, bitstream-based models are suitable for cloud gaming providers as well as network providers to monitor the quality of cloud gaming service. Due to the importance of this type of model for quality monitoring, it is expected to see more efforts in the development of an accurate bitstream-based model for cloud gaming services under different encoding settings. As an example of such efforts, a new work item is established within ITU-T Study Group 12 to develop a bitstream-based model for cloud gaming service, named ITU-T P.BBQCG [132].

[1] https://chromeunboxed.com/stadia-av1-compression-codec.

- *HDR, 4k, and 144Hz*: Some cloud gaming service providers already offer 4k and HDR content, e.g., Stadia. It is expected with the fifth generation technology standard for broadband cellular networks (known as 5G), more bandwidth and lower end-to-end latency are available to the user of the cloud gaming service, allowing them to have an immersive experience with high bitrate demanding technologies. This requires developing a video quality dataset based on the 4K and HDR content and updating the coding impairment accordingly.
- *User Profiling*: One of the significant steps towards a holistic gaming QoE model is to integrate user-related information into the model. Profiling the users of the service according to the level of sensitivity towards certain types of degradation could help in the development of a user-dependent model. Such a profiling method could be done based on demographic information, service usage, and user preferences.

Appendix A
Additional Material Related to Subjective Experiments

In this Appendix, a short overview of the questionnaire, items, and collected demographic information of the interactive dataset are described. The interactive test was conducted with 180 participants. The demographic information collected in the experiment is given in Table A.1. In the following, the items for pre-test, post-game, post-test, and post-condition questionnaire are listed.

Table A.1 Demographic statistics of test participants in interactive dataset

Gender	Female	Male	Transgender	Others	
	73	107	0	0	
Age	18–25	26–30	30–35	35–41	
	32%	31%	35%	2%	
General gaming expertise (beginner—intermediate—expert) [%]					
15.6	11.1	41.7	23.9	7.8	
Experience with used game (beginner—intermediate—expert) [%]					
32.8	27.8	26.7	8.3	4.4	
Hours per week spend on playing video games					
0	0–1	1–5	5–10	10–20	>20
21.7	24.4	27.8	18.9	5.6	1.7
Likes playing (strongly disagree—undecided—strongly Agree) [%]					
1.7	2.2	15.6	52.2	28.3	
Device	PC	console	Smartphone	Others	
	52.2	24.4	22.8	0.6	

S. Zadtootaghaj, *Quality of Experience Modeling for Cloud Gaming Services*,
T-Labs Series in Telecommunication Services,
https://doi.org/10.1007/978-3-030-98249-2

Pre-test Questionnaire

1. *What is your Year of Birth?*
2. *What is your gender?*
    a. Female
    b. Male
    c. Transgender
    d. Prefer not to say
3. *Roughly how many hours per week do you spent on playing video games?*
    a. Between 0 to 1 hours
    b. Between 1 to 5 hours
    c. Between 5 to 10 hours
    d. Between 10 to 20 hours
    e. More than 20 hours
4. *Roughly how often do you play video games in a week?*
    a. Never
    b. Between 1 to 3 times a week
    c. Between 3 to 7 times a week
    d. Between 7 to 14 times a week
    e. More than 14 times week
5. *How would you describe your gaming experience (expertise)?*
    a. 1 – Beginner
    b. 2
    c. 3 – Intermediate
    d. 4
    e. 5 – Expert
6. *I like playing video games.*
    a. 1 - Strongly Disagree
    b. 2 – Disagree
    c. 3 – Undecided
    d. 4 – Agree
    e. 5 - Strongly Agree
7. *On which kind of device do you usually play games?*
    a. PC (Desktop)
    b. Smartphone / Tablet
    c. Console (PlayStation, XBox, ...)
    d. Others

8. *What kind of monitor are you typically using when playing?*
   a. Television (>30″)
   b. Desktop Monitor (>20″)
   c. Laptop (>12″)
   d. Tablet (>8″)
   e. Large Smartphone (>5″)
   f. Small Smartphone (<5″)
   g. Other
9. *How experienced are you in playing the game "[game name]"?*
   a. 1 – Unexperienced
   b. 2
   c. 3 – Intermediate
   d. 4
   e. 5 - Expert

Post-game Questionnaire
The post-game questionnaire covers the following aspects: performance indication (PI), learnability (LE), appeal (AP), and intuitive controls (IC). Component scores are computed as the average value of its items. The used items are summarized in Table A.2 whereas an example of the used 7-point EC-ACR scale for all items is shown in Fig. A.1.

Post-test Questionnaire
For the post-test questionnaire, the following instructions are given to participants: "In the following, we would like you to tell us about your judgement criteria. Please indicate on the scales below, how important in general (not just for this study) the listed aspects are for your rating of the overall quality of your gaming experience."

Table A.2 Overview of items used in the post-game questionnaire

Order	Item text	Item ID
1	I could easily assess how I was performing in the game.	PI1
2	Learning to operate the game is easy for me.	LE1
3	I liked the graphics and images used in the game.	AP1
4	Learning the game controls was easy.	IC1
5	It was clear to me how my performance was going.	PI2
6	It is easy for me to become skillful at using the game.	LE2
7	The game appealed to my visual senses.	AP2
8	The game controls are intuitive.	IC2
9	I was informed about my progress in the game.	PI3
10	I find the game easy to use.	LE3
11	The game was aesthetically appealing.	AP3
12	It was easy to remember the corresponding control.	IC3

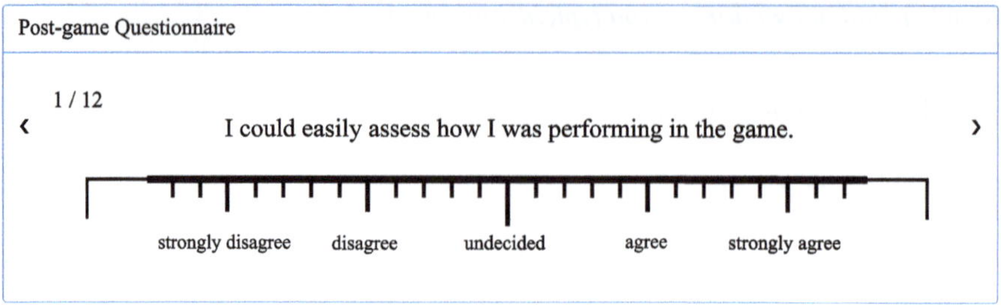

Fig. A.1 Example of item and scales used in the post-game questionnaire

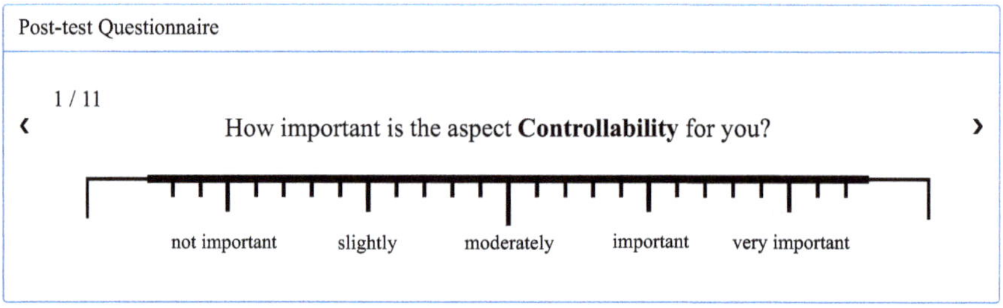

Fig. A.2 Example of item and scales used in the post-test questionnaire

An example of the used 7-point EC-ACR scale for all items is shown in Fig. A.2. The bold written aspect (in the example controllability) was exchanged with: video quality, audio quality, controllability, responsiveness, immediate feedback, video fragmentation, video unclearness, video discontinuity, suboptimal video luminosity, and playing performance. The order of items was randomized and an open question about potential other aspects was added.

Post-condition Questionnaire

Participants are asked to indicate how they felt while playing the game for each of the following items by clicking on the 7-point scale below as explained in the introduction. The questionnaire covers the following aspects:

- Input Quality: Controllability (CN), Responsiveness (RE), Immediate Feedback (IF)
- Output Quality: Audio Quality (AQ), Video Quality (VQ), Video Fragmentation (VF), Video Unclearness (VU), Video Discontinuity (VD), Suboptimal Video Luminosity (VL)
- Player Experience: Immersion (IM), Competency (CO), Negative Affect (NA), Flow (FL), Tension (TE), Positive Affect (PA), Challenge (CH)
- Self-judgement of Playing Performance (PR), and Service Acceptance (AC)

The component scores of each aspect is computed as the average value of its items. The full list of items is summarized in Table A.3 whereas the corresponding rating scales are shown in Figs. A.3, A.4, and A.5.

Table A.3 Overview of items used in the post-game questionnaire

Order	Item text	Item ID	Scale
1	How do you rate the overall quality of your gaming experience?	QOE	1
2	I felt that I had control over my interaction with the system.	CN1	2
3	I noticed a delay between my actions and the outcomes.	RE1	2
4	I felt a sense of control over the game interface and input devices.	CN2	2
5	The responsiveness of my inputs was as I expected.	RE2	2
6	I felt in control of my game actions.	CN3	2
7	I received immediate feedback on my actions.	IF1	2
8	My inputs were applied smoothly.	RE3	2
9	I was notified about my actions immediately.	IF2	2
10	How do you rate the overall audio quality?	AQ	1
11	How do you rate the overall video quality?	VQ	1
12	Fragmentation	VF	3
13	Unclearness	VU	3
14	Discontinuity	VD	3
15	Suboptimal Luminosity	VL	3
16	I found it impressive.	IM1	2
17	I felt successful.	CO1	2
18	I felt bored.	NE1	2
19	It felt like a rich experience.	IM2	2
20	I forgot everything around me.	FL1	2
21	I felt frustrated.	TE1	2
22	I found it tiresome.	NE2	2
23	I felt irritable.	TE2	2
24	I felt skillful.	CO2	2
25	I felt completely absorbed.	FL2	2
26	I felt content.	PO1	2
27	I felt challenged.	CH1	2
28	I had to put a lot of effort into it.	CH2	2
29	I felt good.	PO2	2
30	How do you rate your own playing performance?	PR	1
31	Would you accept using a service under these conditions?	AC	3

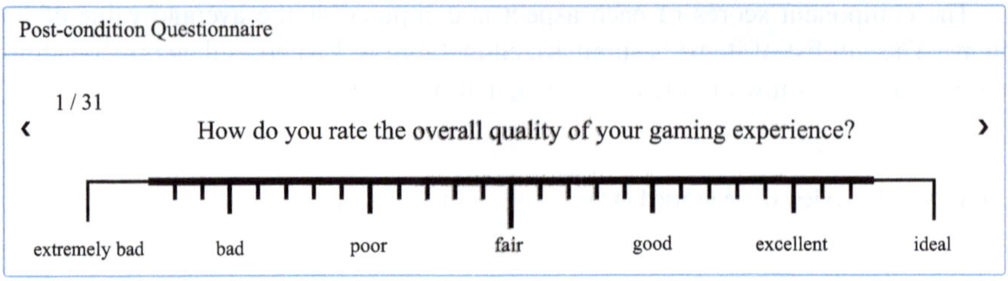

Fig. A.3 Example of the first rating scale type used in the post-condition questionnaire

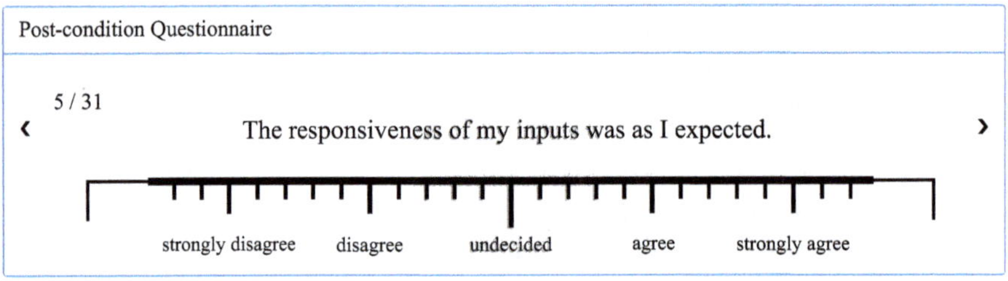

Fig. A.4 Example of the second rating scale type used in the post-condition questionnaire

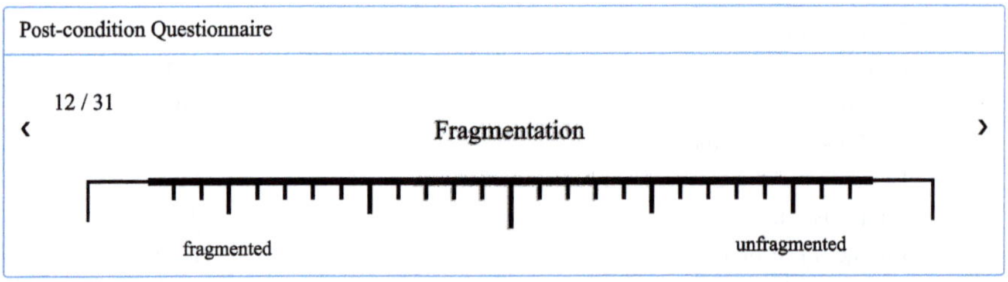

Fig. A.5 Example of the third rating scale type used in the post-condition questionnaire. The 7-point continuous bipolar scale was used which was attached with the following antonym pairs: fragmented and unfragmented (VF), unclear and clear (VU), discontinuous and continuous (VD), suboptimal and optimal (VL), no and yes (AC)

References

1. ITU-T Recommendation G.1032, *Influence Factors on Gaming Quality of Experience*. (International Telecommunication Union, Geneva, 2017)
2. Qualinet White Paper on Definitions of Quality of Experience, *COST Action IC 1003* (2013)
3. ITU-T Recommendation P.910, *Subjective Video Quality Assessment Methods for Multimedia Applications* (International Telecommunication Union, Geneva, 2008)
4. J.M. Juran, F.M. Gryna, R.S. Bingham, *Quality Control Handbook* (McGraw Hill, New York, 1974). No. 658.562 Q-1q
5. K. Ishikawa, *What Is Total Quality Control? The Japanese Way* (Prentice Hall, Hoboken, 1985)
6. ISO 9000, *Quality Management Systems-Fundamentals and Vocabulary* (International Organization for Standardization, Geneva, 2015)
7. W. Robitza, A. Ahmad, P.A. Kara, L. Atzori, M.G. Martini, A. Raake, L. Sun, Challenges of future multimedia QoE monitoring for Internet Service providers, in *Multimedia Tools and Applications*, vol. 76(21) (2017), pp. 22243–22266.
8. ITU-T Recommendation Y.1540, *Internet Protocol Data Communication Service - IP Packet Transfer and Availability Performance Parameters* (International Telecommunication Union, Geneva, 2019)
9. ITU-T Recommendation Y.1541, *Network Performance Objectives for IP-based Services* (International Telecommunication Union, Geneva, 2011)
10. ITU-T Recommendation P.800, *Methods for Subjective Determination of Transmission Quality* (International Telecommunication Union, Geneva, 1996)
11. S. Möller, S. Schmidt, J. Beyer, Gaming taxonomy: an overview of concepts and evaluation methods for computer gaming qoe, in *2013 Fifth International Workshop on Quality of Multimedia Experience (QoMEX)* (IEEE, New York, 2013), pp. 236–241.
12. K. Poels, Y.A. de Kort, W.A. IJsselsteijn, *Game Experience Questionnaire: Development of a Self-Report Measure to Assess the Psychological Impact of Digital Games*. Technische Universiteit Eindhoven, 2007
13. D. Pinelle, N. Wong, T. Stach, Heuristic evaluation for games: usability principles for video game design, in *Proceedings of the SIGCHI Conference on Human Factors in Computing Systems* (2008), pp. 1453–1462
14. ITU-T Recommendation P.809, *Subjective Evaluation Methods for Gaming Quality* (International Telecommunication Union, Geneva, 2018)
15. B.G. Witmer, C.J. Jerome, M.J. Singer, The factor structure of the presence questionnaire. Presence: Teleoper. Virt. Environ. **14**(3), 298–312 (2005)

S. Zadtootaghaj, *Quality of Experience Modeling for Cloud Gaming Services*,
T-Labs Series in Telecommunication Services,
https://doi.org/10.1007/978-3-030-98249-2

16. E. Brown, P. Cairns, A grounded investigation of game immersion, in *CHI'04 Extended Abstracts on Human Factors in Computing Systems* (2004), pp. 1297–1300
17. M. Hassenzahl, User experience (UX) towards an experiential perspective on product quality, in *Proceedings of the 20th Conference on l'Interaction Homme-Machine* (2008), pp. 11–15
18. C. Murphy, Why games work and the science of learning, in *Interservice, Interagency Training, Simulations, and Education Conference*, vol. 21. Citeseer, 2011. [Online]. Available: http://citeseerx.ist.psu.edu/viewdoc/download?doi=10.1.1.467.3543&rep=rep1&type=pdf
19. S. Zadtootaghaj, S. Schmidt, S. Möller, Influence factors on gaming quality of experience (QoE). Available at: https://www.itu.int/md/T17-SG12-C-0104/en, ITU-T Study Group 12, Geneva, ITU-T Contribution C.104, 2017
20. S. Schmidt, S. Zadtootaghaj, S. Möller, Updates on the first draft of influence factors in gaming quality of experience (QoE). Available at: https://www.itu.int/md/T17-SG12-C-0041/en, ITU-T Study Group 12, Geneva, ITU-T Contribution C.41, 2017
21. S. Möller, A. Raake, *Quality of Experience: Advanced Concepts, Applications and Methods* (Springer, New York, 2014)
22. J. Davis, Y.-H. Hsieh, H.-C. Lee, Humans perceive Flicker artifacts at 500 Hz. Sci. Rep. **5**, 7861 (2015)
23. S. Jumisko-Pyykkö, T. Vainio, Framing the context of use for mobile HCI. Int. J. Mob. Hum. Comput. Interact. **2**(4), 1–28 (2010)
24. S. Jumisko-Pyykkö, *User-Centered Quality of Experience and its Evaluation Methods for Mobile Television* (Tampere University of Technology, Tampere, 2011), p. 12
25. H.-E. Yang, C.-C. Wu, K.-C. Wang, An empirical analysis of online game service satisfaction and Loyalty. Exp. Syst. Appl. **36**(2), 1816–1825 (2009)
26. T. Meline, *A Research Primer for Communication Sciences and Disorders* (Allyn & Bacon, Boston, 2009)
27. H. Robin, L. Marc, Z. Robert, A formal approach to game design and game research, in *GDC*, San Jose, 2004
28. S. Schmidt, S. Zadtootaghaj, S. Möller, Towards the delay sensitivity of games: there is more than genres, in *Ninth International Workshop on Quality of Multimedia Experience (QoMEX)* (IEEE, New York, 2017)
29. D. Djaouti, J. Alvarez, J.-P. Jessel, G. Methel, P. Molinier, A gameplay definition through videogame classification. Int. J. Comput. Games Technol. **2008**, 1–7 [Online]. Available http://www.hindawi.com/journals/ijcgt/2008/470350/
30. S. Zadtootaghaj, S. Schmidt, N. Barman, S. Möller, M.G. Martini, A classification of video games based on game characteristics linked to video coding complexity, in *2018 16th Annual Workshop on Network and Systems Support for Games (NetGames)* (IEEE, New York, 2018), pp. 1–6.
31. N. Barman, M.G. Martini, S. Zadtootaghaj, S. Möller, S. Lee, A comparative quality assessment study for gaming and non-gaming videos, in *2018 Tenth International Conference on Quality of Multimedia Experience (QoMEX)* (IEEE, New York, 2018), pp. 1–6.
32. M. Claypool, K. Claypool, Perspectives, frame rates and resolutions: it's all in the game, in *Proceedings of the 4th International Conference on Foundations of Digital Games* (2009), pp. 42–49
33. I. Vilnai-Yavetz, A. Rafaeli, C.S. Yaacov, Instrumentality, aesthetics, and symbolism of office design. Environ. Behav. **37**(4), 533–551 (2005)
34. J. Beyer, V. Miruchna, S. Möller, Assessing the impact of display size, game type, and usage context on mobile gaming QOE, in *2014 Sixth International Workshop on Quality of Multimedia Experience (QoMEX)* (IEEE, New York, 2014), pp. 69–70
35. I. Hupont, J. Gracia, L. Sanagustin, M.A. Gracia, How do new visual immersive systems influence gaming QoE? A use case of serious gaming with oculus rift, in *2015 Seventh International Workshop on Quality of Multimedia Experience (QoMEX)* (IEEE, New York, 2015), pp. 1–6.
36. S. Winkler, Issues in vision modeling for perceptual video quality assessment. Signal Process. **78**(2), 231–252 (1999)

37. P. Quax, A. Beznosyk, W. Vanmontfort, R. Marx, W. Lamotte, An evaluation of the impact of game genre on user experience in cloud gaming, in *2013 IEEE International Games Innovation Conference (IGIC)* (IEEE, New York, 2013), pp. 216–221
38. M. Ries, P. Svoboda, M. Rupp, Empirical study of subjective quality for massive multiplayer games, in *2008 15th International Conference on Systems, Signals and Image Processing* (IEEE, New York, 2008), pp. 181–184
39. Z.-Y. Wen, H.-F. Hsiao, QoE-driven performance analysis of cloud gaming services, in *2014 IEEE 16th International Workshop on Multimedia Signal Processing (MMSP)* (IEEE, New York, 2014), pp. 1–6
40. J. Beyer, R. Varbelow, J.-N. Antons, S. Möller, Using electroencephalography and subjective self-assessment to measure the influence of quality variations in cloud gaming, in *2015 Seventh International Workshop on Quality of Multimedia Experience (QoMEX)* (IEEE, New York, 2015), pp. 1–6
41. K.-T. Chen, Y.-C. Chang, H.-J. Hsu, D.-Y. Chen, C.-Y. Huang, C.-H. Hsu, On the quality of service of cloud gaming systems. IEEE Trans. Multimedia **16**(2), 480–495 (2013)
42. N. Barman, M.G. Martini, H. 264/MPEG-AVC, H. 265/MPEG-HEVC and VP9 codec comparison for live gaming video streaming, in *2017 Ninth International Conference on Quality of Multimedia Experience (QoMEX)* (IEEE, New York, 2017), pp. 1–6
43. M. Claypool, K. Claypool, F. Damaa, The effects of frame rate and resolution on users playing first person shooter games, in *Multimedia Computing and Networking 2006*, vol. 6071 (International Society for Optics and Photonics, Bellingham, 2006), p. 607101
44. ITU-T Recommendation P.911, *Subjective Audiovisual Quality Assessment Methods for Multimedia Applications* (International Telecommunication Union, Geneva, 1998)
45. S. Schmidt, S. Möller, S. Zadtootaghaj, A comparison of interactive and passive quality assessment for gaming research, in *2018 Tenth International Conference on Quality of Multimedia Experience (QoMEX)* (IEEE, New York, 2018), pp. 1–6
46. M. Claypool, Motion and scene complexity for streaming video games, in *Proceedings of the 4th International Conference on Foundations of Digital Games* (2009), pp. 34–41
47. ITU-T Recommendation P.80, *Methods for Subjective Determination of Transmission Quality* (International Telecommunication Union, Geneva, 1993)
48. S. Schmidt, *Assessing the Quality of Experience of Cloud Gaming Services.* PhD Thesis, TU Berlin, 2021
49. ITU-T Recommendation P.918, *Dimension-Based Subjective Quality Evaluation for Video Content* (International Telecommunication Union, Geneva, 2020)
50. S. Zadtootaghaj, S. Schmidt, S.S. Sabet, S. Möller, C. Griwodz, Quality estimation models for gaming video streaming services using perceptual video quality dimensions, in *Proceedings of the 11th ACM Multimedia Systems Conference* (2020), pp. 213–224
51. ITU-T Recommendation G.1072, *Opinion Model Predicting Gaming Quality of Experience for Cloud Gaming Services.* (International Telecommunication Union, Geneva, 2020)
52. P.M. Fitts, The Information capacity of the human motor system in controlling the amplitude of movement. J. Exp. Psychol. **47**(6), 381 (1954)
53. M. Claypool, Game input with delay—Moving target selection with a game controller thumbstick, in *ACM Transactions on Multimedia Computing, Communications, and Applications (TOMM)*, vol. 14(3s) (2018), pp. 1–22
54. M. Claypool, R. Eg, K. Raaen, Modeling user performance for moving target selection with a delayed mouse, in *International Conference on Multimedia Modeling* (Springer, New York, 2017), pp. 226–237
55. M. Long, C. Gutwin, Characterizing and modeling the effects of local latency on game performance and experience, in *Proceedings of the 2018 Annual Symposium on Computer-Human Interaction in Play* (2018), pp. 285–297
56. M. Claypool, D. Finkel, The effects of latency on player performance in cloud-based games, in *2014 13th Annual Workshop on Network and Systems Support for Games* (IEEE, New York, 2014), pp. 1–6

57. C.-Y. Huang, C.-H. Hsu, Y.-C. Chang, K.-T. Chen, GamingAnywhere: an Open cloud gaming system, in *Proceedings of the 4th ACM Multimedia Systems Conference* (2013), pp. 36–47
58. I. Slivar, L. Skorin-Kapov, M. Suznjevic, Cloud gaming QoE models for deriving video encoding adaptation strategies, in *Proceedings of the 7th International Conference on Multimedia Systems* (2016), pp. 1–12
59. S. Zadtootaghaj, S. Schmidt, S. Möller, Modeling gaming QoE: towards the impact of frame rate and bit rate on cloud gaming, in *2018 Tenth International Conference on Quality of Multimedia Experience (QoMEX)* (IEEE, New York, 2018), pp. 1–6
60. N. Barman, S. Schmidt, S. Zadtootaghaj, M.G. Martini, S. Möller, An evaluation of video quality assessment metrics for passive gaming video streaming, in *Proceedings of the 23rd Packet Video Workshop* (ACM, New York, 2018), pp. 7–12
61. N. Barman, S. Zadtootaghaj, S. Schmidt, M.G. Martini, S. Möller, An objective and subjective quality assessment study of passive gaming video streaming. Int. J. Netw. Manage. **30**, e2054 (2018)
62. S. Göring, R.R.R. Rao, A. Raake, Nofu—A lightweight no-reference pixel based video quality model for gaming content, in *2019 Eleventh International Conference on Quality of Multimedia Experience (QoMEX)*, (IEEE, New York, 2019), pp. 1–6
63. N. Barman, E. Jammeh, S.A. Ghorashi, M.G. Martini, No-reference video quality estimation based on machine learning for passive gaming video streaming applications. IEEE Access **7**, 74511–74527 (2019)
64. S. Van Damme, M.T. Vega, J. Heyse, F. De Backere, F. De Turck, A low-complexity psychometric curve-fitting approach for the objective quality assessment of streamed game videos. Signal Process. Image Commun. **88**, 115954 (2020)
65. S. Wang, S. Dey, Cloud mobile gaming: modeling and measuring user experience in mobile wireless networks. ACM SIGMOBILE Mobile Comput. Commun. Rev. **16**(1), 10–21 (2012)
66. ITU-T Recommendation G.107, *The E-model: a Computational Model for Use in Transmission Planning* (International Telecommunication Union, Geneva, 2015)
67. Y.-T. Lee, K.-T. Chen, H.-I. Su, C.-L. Lei, are all games equally cloud-gaming-friendly? An electromyographic approach, in *2012 11th Annual Workshop on Network and Systems Support for Games (NetGames)* (IEEE, New York, 2012), pp. 1–6
68. I. Slivar, M. Suznjevic, L. Skorin-Kapov, M. Matijasevic, Empirical QoE study of in-home streaming of online games, in *2014 13th Annual Workshop on Network and Systems Support for Games* (IEEE, New York, 2014), pp. 1–6
69. H.-J. Hong, C.-F. Hsu, T.-H. Tsai, C.-Y. Huang, K.-T. Chen, C.-H. Hsu, Enabling adaptive cloud gaming in an open-source cloud gaming platform. IEEE Trans. Circ. Syst. Video Technol. **25**(12), 2078–2091 (2015)
70. I. Slivar, M. Suznjevic, L. Skorin-Kapov, The impact of video encoding parameters and game type on qoe for cloud gaming: a case study using the steam platform, in *2015 Seventh International Workshop on Quality of Multimedia Experience (QoMEX)* (IEEE, New York, 2015), pp. 1–6
71. A. Raake, M.-N. Garcia, S. Moller, J. Berger, F. Kling, P. List, J. Johann, C. Heidemann, TV-model: parameter-based prediction of IPTV quality, in *2008 IEEE International Conference on Acoustics, Speech and Signal Processing* (IEEE, New York, 2008), pp. 1149–1152
72. Z. Wang, A.C. Bovik, H.R. Sheikh, E.P. Simoncelli, Image quality assessment: from error visibility to structural similarity. IEEE Trans. Image Process. **13**(4), 600–612 (2004)
73. H.R. Sheikh, A.C. Bovik, Image information and visual quality. IEEE Trans. Image Process. **15**(2), 430–444 (2006)
74. Netflix, VMAF - Video Multi-Method Assessment Fusion https://github.com/Netflix/vmaf [Online: Accessed 2-Oct-2018]
75. R.R. Ramachandra Rao, S. Göring, R. Steger, S. Zadtootaghaj, N. Barman, S. Fremerey, S. Möller, A. Raake, A large-scale evaluation of the bitstream-based video-quality model ITU-T P. 1204.3 on gaming content, in *2020 IEEE 22nd International Workshop on Multimedia Signal Processing (MMSP)*, (IEEE, 2020), pp. 1–6
76. J. Allnatt, *Transmitted-Picture Assessment* (Wiley, Chichester, 1983)

77. N. Barman, S. Zadtootaghaj, S. Schmidt, M.G. Martini, S. Möller, GamingVideoSET: a dataset for gaming video streaming applications, in *2018 16th Annual Workshop on Network and Systems Support for Games (NetGames)* (IEEE, New York, 2018), pp. 1–6
78. ITU-R Recommendation BT-500-19, *Methodology for the Subjective Assessment of the Quality of Television Pictures*. (International Telecommunication Union, Geneva, 2002)
79. D.C. Hoaglin, B. Iglewicz, Fine-tuning some resistant rules for outlier labeling. J. Am. stat. Assoc. **82**(400), 1147–1149 (1987)
80. F. Köster, D. Guse, M. Wältermann, S. Möller, Comparison between the discrete ACR scale and an extended continuous scale for the quality assessment of transmitted speech. *Fortschritte der Akustik, DAGA*, 2015
81. S.S. Sabet, S. Schmidt, S. Zadtootaghaj, C. Griwodz, S. Moller, Delay sensitivity classification of cloud gaming content, in *Proceedings of the 12th ACM International Workshop on Immersive Mixed and Virtual Environment Systems, MMVE '20*, pp. 25–30
82. A. Espen, S.S. Marie, S. Lise, A multidimensional typology of games in *DiGRA Conference*, 2003
83. S. Schmidt, S. Shafiee Sabet, S. Zadtootaghaj, S. Möller, C. Griwodz, N. Barman, M.G. Martini, Proposal of a Content Classification for Cloud Gaming Services. Available at: https://www.itu.int/md/T17-SG12-C-0444/en, ITU-T Study Group 12, Geneva, ITU-T Contribution C.444, 2019
84. M. Claypool, K. Claypool, Latency can kill: precision and deadline in online games, in *Proceedings of the First Annual ACM SIGMM Conference on Multimedia Systems* (ACM, New York, 2010)
85. M. Claypool, K. Claypool, Latency and player actions in online games, in *Communications of the ACM* (ACM, New York, 2006)
86. S. Schmidt, B. Naderi, S.S. Sabet, S. Zadtootaghaj, S. Möller, Assessing interactive gaming quality of experience using a crowdsourcing approach, in *2020 Twelfth International Conference on Quality of Multimedia Experience (QoMEX)* (IEEE, New York, 2020), pp. 1–6
87. ITU-T Recommendation P.1203, *Parametric Bitstream-Based Quality Assessment of Progressive Download and Adaptive Audiovisual Streaming Services Over Reliable Transport* (International Telecommunication Union, Geneva, 2017)
88. ITU-T Recommendation P.1204.3, *Video Quality Assessment of Streaming Services over Reliable Transport for Resolutions Up to 4K with Access to Full Bitstream Information* (International Telecommunication Union, Geneva, 2020)
89. S. Zadtootaghaj, N. Barman, S. Schmidt, M. G. Martini, S. Möller, NR-GVQM: a no reference gaming video quality metric, in *2018 IEEE International Symposium on Multimedia (ISM)* (IEEE, New York, 2018), pp. 131–134
90. S. Zadtootaghaj, N. Barman, R. R. Ramachandra Rao, S. Göring, M.G. Martini, A. Raake, S. Möller, DEMI: deep video quality estimation model using perceptual video quality dimensions, in *2020 IEEE 22nd International Workshop on Multimedia Signal Processing (MMSP)* (IEEE, New York, 2020), pp. 1–6
91. M. Utke, S. Zadtootaghaj, S. Schmidt, S. Bosse, S. Möller, NDNetGaming-development of a no-reference deep CNN for gaming video quality prediction, in *Multimedia Tools and Applications* (2020), pp. 1–23
92. A. Raake, M.-N. Garcia, W. Robitza, P. List, S. Göring, B. Feiten, A bitstream-based, scalable video-quality model for HTTP adaptive streaming: ITU-T P. 1203.1, in *2017 Ninth International Conference on Quality of Multimedia Experience (QoMEX)* (IEEE, New York, 2017), pp. 1–6
93. A.K. Moorthy, A.C. Bovik, A two-step framework for constructing blind image quality indices. IEEE Signal Process. Lett. **17**(5), 513–516 (2010)
94. A. Mittal, R. Soundararajan, A.C. Bovik, Making a "Completely Blind" image quality analyzer. IEEE Signal Process. Lett. **20**(3), 209–212 (2013)
95. A. Mittal, A.K. Moorthy, A.C. Bovik, No-reference image quality assessment in the spatial domain. IEEE Trans. Image Process. **21**(12), 4695–4708 (2012)

96. M. Leszczuk, M. Hanusiak, I. Blanco, A. Dziech, J. Derkacz, E. Wyckens, S. Borer, Key indicators for monitoring of audiovisual quality, in *22nd Signal Processing and Communications Applications Conference (SIU)*, April 2014, pp. 2301–2305
97. M. Leszczuk, M. Hanusiak, M.C.W. Farias, E. Wyckens, G. Heston, *Recent Developments in Visual Quality Monitoring by Key Performance Indicators*, vol. 75, no. 17, Sep 2016, pp. 10745–10767. [Online]. Available: https://doi.org/10.1007/s11042-014-2229-2
98. F. Chollet, *Keras* (2015). https://keras.io
99. O. Russakovsky, J. Deng, H. Su, J. Krause, S. Satheesh, S. Ma, Z. Huang, A. Karpathy, A. Khosla, M. Bernstein, et al., Imagenet large scale visual recognition challenge. Int. J. Comput. Vis. **115**(3), 211–252 (2015)
100. G. Huang, Z. Liu, L. Van Der Maaten, K.Q. Weinberger, Densely connected convolutional networks, in *Proceedings of the IEEE Conference on Computer Vision and Pattern Recognition* (2017), pp. 4700–4708
101. K. He, X. Zhang, S. Ren, J. Sun, Deep residual learning for image recognition, in *Proceedings of the IEEE Conference on Computer Vision and Pattern Recognition* (2016), pp. 770–778
102. F. Chollet, Xception: deep learning with depthwise separable convolutions, in *Proceedings of the IEEE Conference on Computer Vision and Pattern Recognition* (2017), pp. 1251–1258
103. M. Sandler, A. Howard, M. Zhu, A. Zhmoginov, L.-C. Chen, MobileNetV2: inverted residuals and linear bottlenecks, in *The IEEE Conference on Computer Vision and Pattern Recognition (CVPR)*, June 2018
104. T. Judd, K. Ehinger, F. Durand, A. Torralba, Learning to predict where humans look, in *2009 IEEE 12th International Conference on Computer Vision* (IEEE, New York, 2009), pp. 2106–2113
105. ITU-T Recommendation P.913, *Methods for the Subjective Assessment of Video Quality, Audio Quality and Audiovisual Quality of Internet Video and Distribution Quality Television in Any Environment* (International Telecommunication Union, Geneva, 2016)
106. M. Seufert, M. Slanina, S. Egger, M. Kottkamp, To pool or not to pool: a comparison of temporal pooling methods for HTTP adaptive video streaming, in *2013 Fifth International Workshop on Quality of Multimedia Experience (QoMEX)* (IEEE, New York, 2013), pp. 52–57
107. G. Sperling, Temporal and spatial visual masking. I. Masking by impulse Flashes. J. Sci. Arts **55**(5), 541–559 (1965)
108. L.K. Choi, A.C. Bovik, Video quality assessment accounting for temporal visual masking of local flicker. Signal Process. Image Commun. **67**, 182–198 (2018)
109. ITU-T Recommendation P.910, *Subjective Video Quality Assessment Methods for Multimedia Applications* (International Telecommunication Union, Geneva, 2008)
110. C.G. Bampis, Z. Li, I. Katsavounidis, T.-Y. Huang, C. Ekanadham, A.C. Bovik, Towards Perceptually Optimized End-to-end Adaptive Video Streaming, CoRR, vol. abs/1808.03898 (2018). [Online]. Available: https://arxiv.org/abs/1808.03898
111. Netflix Public Dataset, https://github.com/Netflix/vmaf/blob/master/resource/doc/datasets.md. [Online: Accessed 06 Sept 2019]
112. ITU-T Recommendation G.1071, *Opinion Model for Network Planning of Video and Audio Streaming Applications* (International Telecommunication Union, Geneva, 2016)
113. ITU-T Recommendation P.1201.2, *Parametric Non-Intrusive Assessment of Audiovisual Media Streaming Quality - Higher Resolution Application Area* (International Telecommunication Union, Geneva, 2012)
114. N. Venkatanath, D. Praneeth, M.C. Bh, S.S. Channappayya, S.S. Medasani, Blind image quality evaluation using perception based features, in *2015 Twenty First National Conference on Communications (NCC)* (IEEE, New York, 2015), pp. 1–6
115. Multimedia Signal Processing Group (MMSPG, EPFL), VQMT: Video Quality Measurement Tool. http://mmspg.epfl.ch/vqmt [Online: accessed 12 Feb 2018]
116. S. Zadtootaghaj, Performance evaluation of existing quality models and ITU standards on video gaming quality estimation. file:///Users/saman/Downloads/VQEG_2018_123_CGI_Gaming_Video_Quality_Estimation.pdf [Online: accessed 27 Dec 2020]

117. ITU-T Recommendation P.1401, *Methods, Metrics and Procedures for Statistical Evaluation, Qualification and Comparison of Objective Quality Prediction Models* (International Telecommunication Union, Geneva, 2020)
118. H. Talebi, P. Milanfar, Nima: neural image assessment. IEEE Trans. Image Process. **27**(8), 3998–4011 (2018)
119. N.J. Avanaki, S. Zadtootaghaj, N. Barman, S. Schmidt, M.G. Martini, S. Möller, Quality enhancement of gaming content using generative adversarial networks, in *2020 Twelfth International Conference on Quality of Multimedia Experience (QoMEX)* (IEEE, New York, 2020), pp. 1–6
120. N. Barman, S. Schmidt, S. Zadtootaghaj, M.G. Martini, Evaluation of MPEG-5 part 2 (LCEVC) for live gaming video streaming applications, in *Proceedings of the 1st Conference on Mile-High Video* (2022), pp. 108–109
121. K. Simonyan, A. Zisserman, Very Deep Convolutional Networks for Large-Scale Image Recognition. arXiv preprint arXiv:1409.1556 (2014)
122. I. Goodfellow, J. Pouget-Abadie, M. Mirza, B. Xu, D. Warde-Farley, S. Ozair, A. Courville, Y. Bengio, Generative adversarial nets, in *Advances in Neural Information Processing Systems* (2014), pp. 2672–2680
123. Y. Blau, R. Mechrez, R. Timofte, T. Michaeli, L. Zelnik-Manor, The 2018 PIRM challenge on perceptual image super-resolution, in *Proceedings of the European Conference on Computer Vision (ECCV)* (2018), pp. 0–0
124. Z. Zhao, Q. Sun, H. Yang, H. Qiao, Z. Wang, D.O. Wu, Compression artifacts reduction by improved generative adversarial networks. EURASIP J. Image Video Process. **2019**(1), 1–7 (2019)
125. C. Ledig, L. Theis, F. Huszár, J. Caballero, A. Cunningham, A. Acosta, A. Aitken, A. Tejani, J. Totz, Z. Wang, et al., Photo-realistic single image super-resolution using a generative adversarial network, in *Proceedings of the IEEE Conference on Computer Vision and Pattern Recognition* (2017), pp. 4681–4690
126. O. Ronneberger, P. Fischer, T. Brox, U-net: convolutional networks for biomedical image segmentation, in *International Conference on Medical Image Computing and computer-Assisted Intervention* (Springer, New york, 2015), pp. 234–241
127. K. He, X. Zhang, S. Ren, J. Sun, Delving deep into rectifiers: surpassing human-level performance on imagenet classification, in *Proceedings of the IEEE International Conference on Computer Vision* (2015), pp. 1026–1034
128. Z. Cheng, H. Sun, M. Takeuchi, J. Katto, Deep convolutional autoencoder-based lossy image compression, in *2018 Picture Coding Symposium (PCS)* (IEEE, New York, 2018), pp. 253–257
129. G. Meardi, S. Ferrara, L. Ciccarelli, G. Cobianchi, S. Poularakis, F. Maurer, S. Battista, A. Byagowi, MPEG-5 part 2: low complexity enhancement video coding (LCEVC): overview and performance evaluation, in *Applications of Digital Image Processing XLIII*, vol. 11510 (International Society for Optics and Photonics, Bellingham, 2020), p. 115101C
130. E. Abbruzzese, M. Inouye, Cloud Gaming: Enabling a Next Generation Gaming and Streaming Paradigm. Available at: https://www.interdigital.com/download/5fc521858934bf7f7ce2542a, InterDigital, ABI Research, Geneva, Tech. Rep.
131. B. Naderi, S. Möller, Transformation of mean opinion scores to avoid misleading of ranked based statistical techniques, in *2020 Twelfth International Conference on Quality of Multimedia Experience (QoMEX)* (IEEE, New York, 2020), pp. 1–4
132. S. Zadtootaghaj, S. Schmidt, S. Möller, S. Shafiee Sabet, C. Griwodz, N. Barman, M.G. Martini, R.R. Ramachandra Rao, S. Göring, A. Raake, Proposal for new work item P.BBQCG: parametric bitstream-based quality assessment of cloud gaming services. Available at: https://www.itu.int/md/T17-SG12-C-0489/en, ITU-T Study Group 12, Geneva, ITU-T Contribution C.489, 2020

Index

S. Zadtootaghaj, *Quality of Experience Modeling for Cloud Gaming Services*,
T-Labs Series in Telecommunication Services,
https://doi.org/10.1007/978-3-030-98249-2

MIX
Papier aus verantwortungsvollen Quellen
Paper from responsible sources
FSC® C105338

If you have any concerns about our products,
you can contact us on
ProductSafety@springernature.com

In case Publisher is established outside the EU,
the EU authorized representative is:
Springer Nature Customer Service Center GmbH
Europaplatz 3, 69115 Heidelberg, Germany

Printed by Libri Plureos GmbH
in Hamburg, Germany